MOBILE USER INTERFACE ANALYSIS AND DESIGN: A PRACTITIONER'S GUIDE TO DESIGNING USER INTERFACE FOR MOBILE DEVICES

MOBILE USER INTERFACE ANALYSIS AND DESIGN: A PRACTITIONER'S GUIDE TO DESIGNING USER INTERFACE FOR MOBILE DEVICES

HOKYOUNG RYU

Nova Science Publishers, Inc.
New York

Copyright © 2009 by Nova Science Publishers, Inc.

All rights reserved. No part of this book may be reproduced, stored in a retrieval system or transmitted in any form or by any means: electronic, electrostatic, magnetic, tape, mechanical photocopying, recording or otherwise without the written permission of the Publisher.

For permission to use material from this book please contact us:
Telephone 631-231-7269; Fax 631-231-8175
Web Site: http://www.novapublishers.com

NOTICE TO THE READER

The Publisher has taken reasonable care in the preparation of this book, but makes no expressed or implied warranty of any kind and assumes no responsibility for any errors or omissions. No liability is assumed for incidental or consequential damages in connection with or arising out of information contained in this book. The Publisher shall not be liable for any special, consequential, or exemplary damages resulting, in whole or in part, from the readers' use of, or reliance upon, this material.

Independent verification should be sought for any data, advice or recommendations contained in this book. In addition, no responsibility is assumed by the publisher for any injury and/or damage to persons or property arising from any methods, products, instructions, ideas or otherwise contained in this publication.

This publication is designed to provide accurate and authoritative information with regard to the subject matter covered herein. It is sold with the clear understanding that the Publisher is not engaged in rendering legal or any other professional services. If legal or any other expert assistance is required, the services of a competent person should be sought. FROM A DECLARATION OF PARTICIPANTS JOINTLY ADOPTED BY A COMMITTEE OF THE AMERICAN BAR ASSOCIATION AND A COMMITTEE OF PUBLISHERS.

LIBRARY OF CONGRESS CATALOGING-IN-PUBLICATION DATA

Ryu, Hokyoung.
  Mobile user interface analysis and design : a practitioner's guide to designing user interfaces for mobile devices / Hokyoung Ryu.
      p. cm.
    Includes index.
    ISBN 978-1-60692-903-2 (hardcover)
   1. Mobile computing. 2. User interfaces (Computer systems) 3. Wireless communication systems. I. Title.
    QA76.59.R983   2009
    005.4'37--dc22
                                         2008053595

Published by Nova Science Publishers, Inc. ✦ *New York*

To my loving family:

Jihoon, Youngji and Eunsook

Hyosun and Jongnam

CONTENTS

Foreword		ix
Preface		xi
Acknowledgements		xv
Chapter 1	Introduction	1
Chapter 2	Challenges in Mobile User Interface Design	11
Chapter 3	Mobile UI Design Process	25
Chapter 4	Mobile Context Analysis	35
Chapter 5	Mobile Work Analysis	93
Chapter 6	Task-Function Mapping	119
Chapter 7	Action-Effect Design	137
Chapter 8	Information Design	153
Chapter 9	Collective Walkthroughs	167
Chapter 10	Testing with Users	217
Epilogue	The Way Forward	237
References		239
Index		251

FOREWORD

The mobile phones we all now own are digital devices with functionality and computing power that just a few years ago would have been unimaginable for something one can slip into a pocket. In future years, one can only expect this computing power to carry on growing exponentially, but can the functionality continue to grow in the same way? In this book, Hokyoung Ryu argues that we may have reached a limit to what can be crammed into a single device. This limit is not one imposed by the computing power of the device, rather by human ability to access all this functionality using a small device.

Hokyoung Ryu's essential point is that designing for a mobile device to fit in someone's pocket is not the same as designing for a PC on someone's office desk. The functionality accessible via a PC is staggering, but people expect to be trained to access and use it. They do not expect to have to take a course to learn what their phone can do. Further, the task of the designer of a mobile device is much harder than that of someone designing for a PC. Having a full-size keyboard, mouse and above all, a large high-definition screen makes it much easier to organise material in a way that is self-explanatory. Importantly, there are also well-established conventions for interaction. The user of a PC knows how to interact with a menu, a dialogue box or a scroll bar whether the operating system is Windows, Mac or Linux. The designer of a mobile device, on the other hand, is faced with a tiny display and limited input precision. In addition, there is little consistency between interaction methods across devices.

Hokyoung Ryu is right in stating that future mobile devices will only come fully into their own when designers start to focus more clearly on what users want to do with them. This means that they need to be more specialised. I personally believe that Norman's vision of a wide variety of 'information appliances," each carefully focussed on a particular well-understood set of human tasks, is the only way forward for future mobile devices. Systems delivering GPS navigation for motorists are a successful example of such a specialisation.

This book describes how to achieve the user-centred design of mobile technology. It starts with practical methods for the analysis of the human needs one is aiming to satisfy with a putative device. It then moves through to techniques for design and evaluation. Hokyoung Ryu rightly sees evaluation as a natural part of design, a critical element in the process of getting ideas by the application of which a design can be improved. The book provides analytic and empirical methods for doing this.

This is essentially a practical book for designers, filled with procedures and examples that they can use in their daily work. I hope designers will read it and put it to good use. The possibilities are considerable. We need usable and useful mobile devices, but as long as designers continue to imagine that they are just miniature PCs, they are not going achieve this potential.

— Professor Andrew Monk
Centre for Usable Home Technology
University of York, UK

PREFACE

DESIGNING INTERFACES FOR MOBILE DEVICES SHOULD BE NOVEL EXERCISES AGAINST DESKTOP USER INTERFACES

We are confident that new mobile devices – from smart phones and PDAs (personal digital assistants) to Tablet PCs and beyond – will begin to change the way we used to live without them. More importantly, we are all well aware that they will bring about fundamental advances in the way we work. However, we must be aware that mobile devices for their own sake will never be powerful tools unless they make sense in the everyday life of a user. We must constantly make sure that the tools we provide are useful, usable, affordable and comprehensible. However, it is evident that this dream has not yet been achieved. Stories abound of successful mobile devices – usable or otherwise – entering the marketplace, and many more are still coming up at this time of writing.

This book is the articulation of my years of practical experience, teaching and research in mobile user interface design. I frequently encounter frustrated users who have found mobile interfaces difficult to understand and use, and even mobile interface practitioners have wondered about how to make themselves open to such a perfect design exercise. This book is the outcome of experience beyond my own. Many colleagues, both those with whom I have worked directly and those whose wisdom I have read and borrowed, have added to this humble understanding. However, I do not intend to suggest that you use this book as a cookbook for mobile user interface design; rather, I hope that you find the ideas, stories, and guidelines in this book stimulating as you design your mobile interfaces. Moreover, I hope that the experiences and techniques presented in this book provide you with sufficient detail to get started on your own excursions into your mobile user interface design world.

Contrary to the view that this book may guide practitioners to kick off mobile user interface design—gathering information that will help you develop a mobile user interface that users will want to use and be able to use easily and efficiently—researchers who are interested in a new interaction design framework will find philosophies for thought and techniques to exploit the potential of cognitive psychology; therefore, a special effort has been made to include relevant examples for further thought.

Nonetheless, based on what I know, the main route of this book remains intact, i.e., to help you to develop mobile user interfaces. It is all about that!

THE CHALLENGES

The increasing spread of mobile devices is dramatically affecting people's daily lives. They not only increase the pace and efficiency of everyday life, but also allow more flexibility at business and professional levels.

Arguably, mobile technologies, particularly the increasingly sophisticated mobile phone (blurring the boundaries between communication and computation), combine both ubiquity and utility. This phenomenon has given rise to opportunities to employ mobile technologies more broadly than just as communication tools. For instance, learning design with mobile technologies has been, to some extent, a successful technological and scientific undertaking, helping us broaden the arenas of the educational sector in ways that no one could otherwise have anticipated a decade ago. Of course, we cannot predict its future; however, the progress of the past decade highlights specific current challenges.

To the extent that the success of mobile phones points out that we must develop a better understanding of how mobile technologies can be used to enhance various user experiences, empower the user with the knowledge and ability to self-manage, we must learn how these technologies can improve quality of life across a spectrum of contexts while containing costs and stimulating demand for mobile services.

As the technologies that may support new mobile devices continue to evolve, this field will become increasingly more challenging as new opportunities emerge. For instance, how to effectively take the user into account within emerging mobile environments has formed a persistent theme in the academic field of human-computer interaction (HCI). In contrast, much of the practitioner's perspective on mobile applications has been what kind of products and content can facilitate the uptake of this new mobile environment. To achieve significant outcomes from this research that both deliver technological solutions and enhance the usability and sustainability of the technologies, this book aims to draw together expertise from a range of my academic and industrial experience. Although this theme has been widely discussed in the existing texts, there are still few available empirical, exploratory or systematic design processes identified as crucial.

OVERVIEW OF THE BOOK

This book is intended to discuss the latest mobile interface design beyond the desktop interface design environment, an area of research that is increasingly seeing new developments and techniques in both the academic and practitioner's fields. It comprises my many years of industrial experience and research in the field of mobile user interfaces. One purpose of the book is to disseminate thoughts about the challenges and practical experience of the design of mobile interfaces, current developments in mobile product experiences in academia and industry, current methods and approaches to mobile interface development, and the current economic and social context of mobile interface development. More importantly, a key aim of this book is to explore the technical aspects of mobile user interface design, where we need to systematically take into account user interactions, activities and the completely renewed social and cultural environments that mobile environments can integrate with and that technologies are now capable of delivering.

The chapters are organized in the book along some general dimensions of user interface design process: *understanding work contexts, designing tasks, putative design exercises,* and *testing*. Before this level of description, some theoretical foundations for designing mobile user experiences are first described in Chapters 2 and 3. This will help the reader understand the structure of this book. We then focus on work context analysis with mobile environments in Chapter 4, work analysis in Chapter 5, and iterative design exercises in Chapters 6 through 9. Finally, Chapter 10 addresses usability testing with users in developing mobile user interfaces. This organization will hopefully assist the reader in seeing various perspectives of current mobile user interface design projects, but may be regarded as somewhat idealized. However, we see the important connecting factor between the chapters as their focus on common themes and arguments in a holistic and systematic design process.

TOWARDS A SOLUTION

Mobile user interface analysis and design is a relatively new research area in human-computer interaction. There is an increasing demand for tools and techniques, but perhaps less enthusiasm or support for practitioners to have the opportunity to fully articulate the relationships among these tools, techniques and underlying HCI theories. Therefore, a comprehensive volume covering current trends, technologies and techniques in mobile user interface design is necessary. In this prism, I believe that this book will be a timely publication for both researchers and practitioners who are interested in the design and development of future mobile products. However, this is, of course, a collection of my humble thoughts on related topics. Readers of the book should thus not feel constrained by the order of the chapters or the structure of the book for their interests. Obviously, I have arranged the material in an order that makes sense to me by taking my own mobile UI design process into consideration, trying wherever possible to closely locate readings and thoughts that speak to the same or related issues, but many different arrangements are possible, and these reinterpretations may suggest other solutions to the future challenges of mobile user interface analysis and design. Finally, I must also acknowledge that this book does not merely communicate with the reader regarding mobile user interface issues; it endeavors to make the most of my words as an externalization of my thoughts, so I know it may be very humble!

ACKNOWLEDGEMENTS

It is difficult to remember all of the many people who have helped to make this book a reality during the writing and preparation of this monograph. I am especially indebted to the reviewers who commented at length on earlier versions of the manuscript. In addition, I am in indebted both to Professor Andrew Monk and Professor Wan Chul Yoon, who have inspired me to complete this humble work as a tangible outcome. I am also grateful to Professor Tony Norris for his leadership at the Centre for Mobile Computing at Massey University, within which my research into mobile user interface analysis and design is based. In closing, I would like to warmly thank all of the mobile user interface researchers, whose work has been borrowed in this book, for their valuable insights and studies which have served to galvanize this book.

Special thanks also go to the publishing team at Nova Science Publishers, Inc., whose contributions throughout the whole process from inception of the initial idea to final publication have been invaluable.

Finally, there is my family—Jihoon, Youngji, and Eunsook—to whom any formal expression of thanks seems inadequate.

— Hokyoung Ryu
Auckland, New Zealand

Chapter 1

INTRODUCTION

There is little doubt that mobile technologies are among the defining technological transformations of the late twentieth and early twenty-first centuries. They have allowed us to explore new opportunities around "m-" neologisms such as m-office, m-government, m-commerce, m-health, m-learning, and so forth. Indeed, the prevalent mobile paradigm often twists designers' arms to fashion new ways in which people fulfill their daily lives, effectively and pleasantly, as their pivotal intent. However, we are well aware that the success of such new mobile services is highly reliant upon affordable and effective applications that are well matched to the needs of the user contexts and environments.

Nonetheless, it is still common to believe that simply adding new features or the catchy phrase '*going mobile*' will ensure commercial success under the all-but-mobile paradigm. Look at your own mobile phone; it may include a dozen otherwise single devices, such as camera, MP3 player, global positioning system (GPS) and, further, a tiny office tool (calendar, scheduler and so forth). While these features in your mobile phone would be of great power in certain cases, perhaps they are better known as *feature fatigue* in that they will eventually turn your mobile phone into a disadvantage (e.g., higher cost, poor usability and so forth). It is only a tactic that works superbly in distinguishing one mobile phone against others, but which generally undermines actual users needs.

In fact, this technology-oriented approach overlooks the possibilities for embodying the fundamental usability concept. We are all aware of systems that add new features with every new version. Such features may not relate well, if at all, to user needs. They are often added simply because they can be added. Indeed, adding new features can provide multiple options for carrying out tasks which are simply not as imminently needed. Often many new users are unaware of many of the functions that a mobile device and its applications can offer them, arguably saying that many smart functions are never used by them. Paradoxically speaking, adding new functionality may even disempower or confuse experienced users. New users may also become confused by the emerging complexity of mobile devices and be discouraged from making the effort required to learn to use them properly and efficiently.

The world of technology is becoming more competitive. It is no longer sufficient to use the number of functionalities of the technology and to add as many features as possible in order to simply hold out their virtues as an innovative offering. This is where mobile user interface analysis and design is becoming more and more important: as a means of achieving more usable and pleasing products against others.

In the modern world, the focus is now on what people can do with the mobile technology, and this is what really counts. In so doing, we need to have a radically different approach to mobile interface design and to the development of design skills. The aim of this book is thus to return to the fundamental area of mobile user interface analysis and design, and to prepare you (mobile system designers or interface practitioners) for more advanced designing of future mobile products. The key to this commitment is to understand better the designs that people need and an appropriate design process – so that people's needs will be met by new system designs in the future.

A COMMON DENOMINATOR OF DESIGN

A good starting point for any designing activities is not to consider what the technology can do. Instead, we must start with an appreciation of potential users and the types of tasks they will need or want to do, which has long been incubated in *user-centered design* (UCD) and *task-based design*. Therefore, designers must have some insight into the needs of their users. This insight will vary from a simple appreciation of the preferences and/or interests of the users to *user models* which capture their skills and abilities in detail (Bushey, Deelman, and Mauney, 2006). Though the benefits of UCD are clear, i.e., the creation of a product that fits users' needs and capabilities, rather than being defined by features on the basis that the designer would want to use them, it does not come easily. There is still a significant drawback in that we think only of a group of representative users and not the whole range of users (or population), which is barely possible; as a consequence, we may end up with a product that is not appropriate for this wider audience market. One common denominator of all design activities is thus to pinpoint *"What do the users really want to have?"* This is not at all an easy job to accomplish; at most a handful of design processes and designers' own individual experiences will be counted among the success.

Rather than understanding the functional needs of the users, it is also very important to appreciate the current *work practices* that people are currently carrying out with (or without) the system given. Often a new system design comes from a new idea thought up by a designer (only because they believe it is "cool"); thus, it will not necessarily benefit users if it does not reflect their current work practices (e.g., a chopstick for eating soup). Of course, there is nothing wrong with this; however, one of the widely-noted features of UCD is that your design will be ruggedized if you pay more attention to the needs and work practices of the users rather than designing it for yourselves. Perhaps a new design starts with a realization that existing systems have design or functional problems. That, too, can be a useful starting point, particularly if it leads you to good ideas about possible solutions. However, as Moggridge (2007) stated, the danger here is that, in starting with other systems, you may tend to repeat the assumptions and errors of those designs. Cool design ideas have potential value, but we run the risk of sticking to our own original thoughts even when our users would prefer other options. This is why it is so crucial to go back to the current work practices and their intrinsic objectives of our intended user groups, e.g., rather than a hollow chopstick for eating soup, a spoon is better than the cool chopstick design.

In user-centered design, evaluation and testing is a sufficient and necessary practice and should be an early component of the whole design process, as opposed to just a one-off performance with a working prototype at the end of the stage. A good designer will iterate this *design-evaluation cycle* through the different design components, assessing his or her own design intent if we are to remain close to what users actually want or need to do. Here, every design effort seems geared toward a system being easy to use. However, it is very difficult to define what "ease of use" means, and to show that one system is easier to use than another. This is because ease of use is "qualitative", which is really hard to measure. Here, another common denominator in mistakes ordinarily made by designers is confusing *usefulness* for *usability*. Consequently, when it comes to evaluating their design, they tend to over-estimate the potential of useful functions in their design rather than its *usable-ness*. For something to be useful, it must allow the user to accomplish a desired task. For something to be usable, it must allow the user to accomplish the desired task easily and enjoyably.

TUPPENCE WORTH

Having considered the common denominators of general design activities, this book discusses a practical methodology appropriate for mobile interface design, which we define as a *systematic design process* whereby mobile user interface designers plan and specify their user interface. Indeed, a surge in methods and techniques of doing UCD has added more burdens to the designer than necessary, but it is the author's belief that there is still a lack of exploration of how to overarch them in a single effective design process from start to end.

The interface designer's objective is to make devices easier and more enjoyable to use, and to make it possible for people to perform tasks which they might not otherwise be able to accomplish. Within this objective in mind, I will propose a guiding interface design process and offer a practical methodology for applying this process with hands-on experiences. This approach is worth considering because it is empirically based, potentially formal, and in concert with current thinking of information design based on cognitive science. It is, of course, up to the reader to decide whether or not the arguments and methodology presented here are sufficient to justify such optimism.

I am well aware that design itself is inherently creative and unpredictable, so a predefined style for design processes might keep us from creating new and novel ideas. However, I personally believe that mobile user interface design must be somewhat systematic, blended with a thorough knowledge of technical feasibility and a mystical aesthetic and cognitive sense of what attracts users. Otherwise, the characteristics that make a mobile device less attractive are its functionality and usability. Here, it is thus worth returning to Carroll and Rosson's (1985) view of design activities:

- Design is a *process*: it is not a state and it cannot be adequately represented statically.
- The design process is *nonhierarchical*; it is neither strictly bottom-up nor strictly top-down.
- The process is *radically destructive* (at least, not constructive); it involves the development of partial and interim solutions that may ultimately play no role in the final design.
- Design intrinsically involves the *discovery of new goals*.

While the platform, user context, devices and technologies involved in a particular mobile device may be different from similar desktop computer applications, the fundamental product design and development processes may remain intact. But in the mobile domain, there can also be discipline, refined techniques, wrong and better methods, and measures of success. The purpose of this book is thus to give mobile product designers, user interface developers, usability professionals, and other product development professionals the tools they need to make the transition into the mobile domain. This book is not about technology or specific design guidelines for particular devices; instead, it aims to cover mobile users and their contexts, heavily leaning toward principles of human-computer interaction, cognitive psychology, usability, and mobile user interface analysis and design. In effect, the challenge of this book is to get you thinking critically and inventively about the ways in which people use mobile systems to support what they want to achieve.

CHANGING FACES OF MOBILE USER INTERFACE DESIGN

In defining the scope of mobile user interface, we do not have any intention to limit ourselves to its focus on only mobile phone interfaces, though many of the examples described in this book capture the design exercises of mobile phones, partly because it is a very common device in our daily lives, but mostly because its knock-on effects on other mobile interfaces is tremendous. In this regard, it may be a good starting point to see how the mobile phone user interface has evolved, by which we will achieve a sense of a certain level of design basics or implications for our next mobile user interface design.

Many of today's high-powered mobile phones come equipped with text messaging, games, video-capturing, cameras and music-playing capabilities. While these applications have become an integral part of the current consumer's experience, the primary feature users demand from phones is simply being able to make a voice call reliably to anyone from anywhere. To meet this fundamental need (although this beauty has been almost forgotten by now), there was Motorola™ DynaTAC 8000X, the first portable mobile phone in the world, shown in Figure 1.1. It weighed 28 ounces and measured 13 inches × 1.75 inches × 3.5 inches, which was quite unwieldy by today's standards, but small enough to be carried. In addition to the typical 12 numeric keys, it had nine additional special keys (e.g., Clr [Clear] and Fcn [Function]), which were quite similar to the landline telephone at that time. Compared to a modern mobile phone, e.g., Nokia™ N70 at 109mm × 53 mm× 24 mm and weighing 126 grams, it was a huge and bulky phone, but indeed it readily captured the future mobile trend, or people's needs, along with being witnessed in the subsequent technological transformation of 1990s.

As the mature technologies that supported new mobile devices continued to evolve in the 1990s, mobile technologies—particularly the increasingly sophisticated mobile devices (blurring the boundaries between communication and computation)—combine both *ubiquity* and *utility*, such as the tiny personal computer, Apple™ Newton MessagePad, which was the first personal digital assistant (PDA), as shown in Figure 1.2. Newton MessagePad had proposed a complete reinvention of the personal computing paradigm, previously perceived as computers on the desktop rather than in the pocket. One widely-noted feature allowed text to be entered by tapping with a stylus pen on a small on-screen pop-up QWERTY virtual keyboard, along with accepting free-hand writing recognition. As such, it radically overcame one of the physical form factors in every mobile user interface design, i.e., the physical keypad on the front of the product. This design freedom is further exploited in the modern touch-sensitive mobile phone design, such as Apple™ iPhone.

Figure 1.1. 1972 brings the first mobile phone, Motorola™'s DynaTAC. (Downloaded from Motorola.com.)

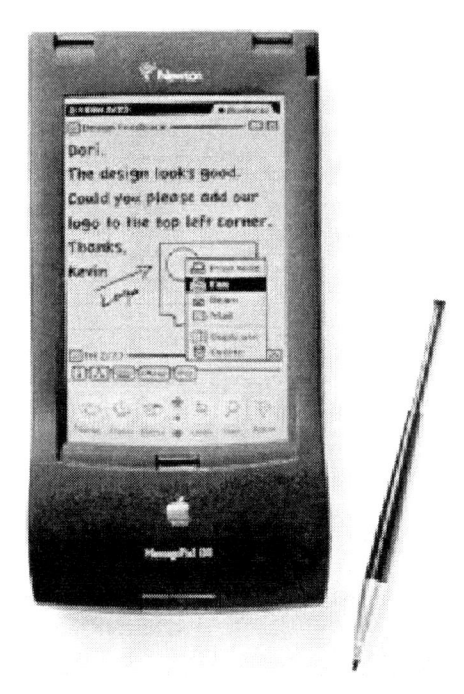

Figure 1.2. 1993 sees the arrival of Apple™ Newton MessagePad100. (Downloaded from Apple.com.)

Coming through these notably marked technical advances in the 1990s, a logical design intent thought of by Nokia™ was to put together the phone and computer functionality, i.e., smart phone[1]. To be fair, in addition to serving as a mobile phone, the Nokia™ 9000 also offered the following features: a calendar; an address book; a tiny version of office applications such as note pad, email, sending and receiving faxes; and games. As shown in Figure 1.3, the device is fixed via a hinge (one side is a PDA, and the other adapts a phone design).

Figure 1.3. 1996 sees the launch of the first 'smart phone'. The Nokia™ 9000 communicator offered Internet access and the ability to send and receive emails, faxes and SMS messages. (Downloaded from Nokia.com.)

More recently, introduction of new mobile phone products has further served to spur competition against the previously major vendors in the mobile phone market, leading to many nip-and-tucks on mobile phone interfaces. Providing new services (e.g., watching TV on the phone—see Figure 1.5), or at least a distinguishing fashionable item, is thus indispensible to winning new market shares. In this sense, Korean mobile phone manufacturers Samsung™ and LG™, which have channeled key resources into new user interfaces or aesthetic designs and new services ahead of its rivals, outperformed the traditional market leaders Nokia™ and Motorola™. For instance, as shown in Figure 1.4, Samsung™ SCH-S310, a.k.a. *motion phone*, has a three-axis accelerometer and three-axis magnetic sensor that give it a range of motion-detection capabilities. The company claimed that the phone could detect motion in three dimensions as well as detect the speed of movements. One of the S310's coolest tricks allows users to write numbers in the air that the phone can then dial, a remarkably new input method beyond the traditional key-in method.

Now, it is highly expected that a further heating up of competition among mobile Internet services and new touch-based interfaces will take place, aimed at bringing in younger and professional users. For instance, recently released to the European market, the LG™ OZ phone is designed for 3G (third generation) mobile Internet access with a stylus pen or QWERTY keypad and an automatic portrait-landscape convertible screen, for better

---

[1] In the literature, the definition of smart phone depends on the author's view, but it is often expressed as a mobile phone offering advanced capabilities beyond a typical mobile phone, often with PC-like functionality.

readability on a small screen, giving users an Internet experience similar to that from a desktop computer.

Figure 1.4. 2005 sees Samsung™ SCHS310 Motion Phone, which allows users to write numbers in the air. (Downloaded from Samsungmobile.com.)

Figure 1.5. DMB (digital multimedia broadcasting) launches in Korea. (Downloaded from lge.com.)

In 2007, Apple™ iPhone (Figure 1.6) created quite a sensation before and after it hit the market. None before Apple™ had actually introduced a fully touch-sensitive user interface with powerful impression in one neatly-sized package. The bigger screen allowed better video quality and more workspace. The Internet browsing capability is another important feature in the iPhone. But the primary function of the iPhone is still checking emails[2], which implies that it occupies a niche market over the other major mobile phone manufacturers.

Figure 1.6. 2007 sees a fabulous Apple™ iPhone. (Downloaded from apple.com.)

Figure 1.7. 2008 sees the Samsung™ haptic phone (SCH-W420). (Downloaded from Samsungmobile.com.)

To compete against Apple™ iPhone, Samsung™'s latest SCH-W420 (a.k.a. *haptic phone*) user interface, as shown in Figure 1.7, is widget-based like the competitor, but further supports easy dragging and dropping of favourite applications or tools on the screen, which is similar to what we have experienced in the desktop computing environment, i.e., *direct manipulation* with *drag-and-drop* interaction. Remarkably, this phone includes several haptic feedback or vibrations when one touches the screen, to compromise the critical limitation of touch-based interaction (i.e., no proper tactile feedback)[3].

Please note that this section has intentionally overlooked (for easy exposition) the advent of several important mobile devices (e.g., MP3 players, personal multimedia player, portable game consoles and so forth), believing that the same account of the mobile phone evolution,

---

[2] Some surveys showed over 77% of iPhone owners mostly use it for email, browsing and downloading rather than phone calls.
[3] http://www.youtube.com/watch?v=UfaGzVuhgxM, for a video clip access.

i.e., from a simple to a more sophisticated and complex mobile device, would be applied in the exact same way.

This book is intended to discuss the latest mobile interface design beyond the desktop-based interface design environment, an area of research that is increasingly seeing new developments and techniques in both the academic and practitioner's fields. I would like to draw together the strands of this book by stating a range of aspects that seem to flow from the analysis to the design, hopefully covering substantial issues associated with the interface design for mobile devices.

The rest of this book is about looking at the fundamental areas of mobile user interface analysis and design, and to provide you with a design process that understands better the mobile designs that people need. A brief overview of the rest of the book is given here.

- *Chapter 2:* Mobile interface design may not be the same as our prior interface design experience in designing desktop computer applications. Simply too many functions would not be best-suited for a small device. The discussion sets out what aspects and characteristics of mobile interfaces would make mobile interface design differ from conventional interface design (e.g., desktop applications) and makes a case for why we have to see the difference critically.
- *Chapter 3:* This chapter covers an iterative mobile interface design process from start to finish, tailored from the traditional user-centred design process. It proposes a putative mobile interface design process.
- *Chapter 4:* User experiences and expectations are the foundations on which users relate to the rest of the world. This chapter provides some real-world examples and defines how to collect users' experiences and expectations and how to exploit them to make a better mobile interface. Also, user interface design should begin with the study of general characteristics of human cognition and perception. An *information-processing model* of human memory is discussed, with implications for mobile user interface design. Human and computer strengths and weakness are also described.
- *Chapter 5:* This chapter discusses the three strands of work analysis methods, for *classifying, decomposing* and *cognitively designing new task practices.* Then interface metaphors are defined and examples are discussed.
- *Chapter 6:* Simply transferring a desktop-based application to the mobile environment always results in a suboptimal mobile experience, partly because of the physical constraints, and mostly because mobile users are primarily interacting with the world through a mobile device and/or with other persons. This chapter discusses how to couple user tasks with the corresponding functions in the mobile device, sketching out a first interface design.
- *Chapter 7:* For updating the first prototype, we have a need to direct our focus onto features of the interaction that are directly affected by system implementation, i.e., input (or action) or output (or system feedback, system effect). This chapter sets out a way to build a consistent *action-to-effect design* in our putative design exercises (introducing a second design prototype).
- *Chapter 8:* The success of a mobile interface, nowadays, is indicative of the power in a more visual and graphical manner. A visual information presentation and effective organization of information would dramatically help our users. In our final design exercise, both screen and navigation design issues are discussed.

- *Chapter 9:* As we have design prototypes from every design exercise, they are evaluated before performing a full-scale usability test, partly because this would save our tight budget and time in the testing stage, but mostly because it would justify our design decisions to be propagated into our iterative design practices. This chapter proposes three methods that are able to assess *congruence, consistency* and *coherence*.
- *Chapter 10:* Usability must be incorporated into a mobile product; therefore, usability testing methods are defined in this chapter. Several types of usability tests are also discussed, based on my many years of academic and industrial experiences towards a pragmatic usability testing solution.

Chapter 2

CHALLENGES IN MOBILE USER INTERFACE DESIGN

David Liddle, in the book entitled *Designing Interactions* (Moggridge, 2007), described leading the team that developed the Star graphical user interface (which is precursor of Apple™ OS and Windows™) and rightly talked about three stages in the development of a technology. The first stage is the *enthusiast* stage, where people do not care whether the technology is easy or hard to use because they are so excited by the technology itself (perhaps like Motorola™ DynaTAC in 1972). People want it without considering whether or not it is difficult to use. Because it is so cumbersome, few would want it, and furthermore it is very expensive. The second stage is the *professional* stage, where those who use the technology are often not those who buy it. A purchasing department usually chooses office computers, and the purchasers do not care much about the difficulty and learning aspects. Instead, they care about performance and productivity (e.g., most early adopters of mobile phones were mobile workers who desperately needed to carry the mobile device to be in touch with their office). At this stage, some people even have a vested interest in the difficult technology because they are selling their ability to use it; the harder it is, the more valuable their skills are. The third stage Liddle terms the *consumer* stage. People now are less interested in the technology itself than what it can do for them. So, if it is hard to use, they will not buy it. (We, by today's standards, all appreciate a mobile phone with useful functions that are simple and easy to use.)

Having reflected on the three stages in the evolution of a technology, AvantGo™ Mobile Lifestyle Survey (2004) shows a conflict of interest in mobile systems design. According to the survey, on the one hand, many mobile consumers want a device that would have the standard PDA features as well as the ability to store and play music, like Apple™ iPod, and with a larger screen than standard mobile phones. On the other hand, at the same time, they want to have a compact mobile device appropriate in size for their pocket (i.e., that they can always carry), and at an affordable price. This self-contradictory interest in what the mobile devices can do for them would generally make designers bring their own common sense to conclude that there is no guarantee at all about what would make our users more comfortable. This issue has gained great attention, but a specific answer has not yet been introduced.

The sole purpose of mobile interface design is thus to provide *users* with *useful* functions in *usable* ways, a.k.a. 3Us. The only test permitting us to judge it good or bad is the comparison between what designers have and what users would like to have in their hands.

This book is about mobile devices and the factors that you should be aware of if you are interested in carrying out research in the area known as mobile user interface analysis and design. In so doing, first we need to see what aspects and characteristics of mobile user interface would differ from the conventional user interface design, and understand why we have to see the differences, if any, critically. Therefore, firstly, in this chapter we aim to introduce you to what aspects should be further considered in mobile user interface design, briefly summarizing four major issues (feature fatigue, paradox of mobile user needs, design freedom, and killer application), building upon the accounts above.

CROWDS OF FEATURES – FEATURE FATIGUE

Mobile technology is supposed to make our lives easier, allowing us to do things more quickly and efficiently. The valuable features of mobile technology have been well captured by Ballard's book (2007), entitled *Designing the Mobile User Experience*, as follows:

- *Mobile* – the user is mobile, not static when using the device.
- *Personal* – the mobile device generally belongs to only one person.
- *Communicative* – the mobile device can send and receive messages of various forms and connect with the network in various ways.
- *Portable* – one of the remarkable differences in mobile capabilities is that people always carry this mobile device, dominating the difference between the desktop-based interface and mobile user interface. That is, a user, using any of a set of mobile devices, could be riding a train, sitting in a meeting, walking down the street, driving a car, and focusing on other tasks.
- *Wakable* – the device can be awakened quickly by either the user or the network. For instance, a mobile phone will receive a text message even when in its sleep, or standby, state. Note that most computers, if they are asleep, cannot communicate with the network. This unique service represents social connection, along with the fact that users always carry the device.

But too often these promising mobile features make things much harder, leaving us with too many features in a single product. Features such as photo-taking, audio or video recording, video calling, Wi-Fi network access, media players and so forth are now perceived as the basic commodities of any mobile device, and further, other sensors are commonly embedded without any reservation: touch-sensitive screen entry and hearing for audible cues, tones, speech input or output, temperature sensitivity and velocity and so forth. This complexity costs users time and extra money to afford the product, and the features would not ensure the users' significant needs; rather, they require designers' efforts to make the complex features better.

Why does this happen? The so-called *internal-audience* problem is proposed: the people who design and sell products are not the ones who buy and use them, and what engineers and marketers think is important is not necessarily what is best for users. Quite often, designers love to add or enhance the capabilities of a product with nascent technologies available, believing that these are what the users want, giving users more control and more degree-of-freedom and options. However, lots of added features are not necessarily the ones users want.

Consider Figure 2.1. As one can see from the trends in a series of products, it is common that more functions or options make a product less usable. However, always paradoxically, we fancy more functional products when we decide to acquire the product.

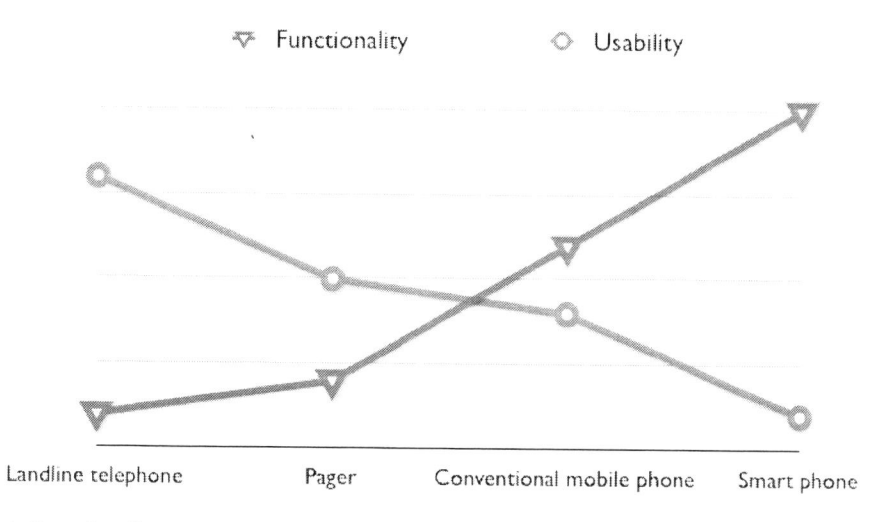

Figure 2.1. Functionality and usability: as more functions are added, usability decreases.

You might simply think, then, that designers should avoid extra options or functions by just focusing on what users really want. But that is where the problem normally arises. Firstly, discarding new design ideas into which designers have invested their creativity is barely possible; and secondly, the overloaded features make buyers (not users) more attracted to the product. For the former, out of good will, the designer should accept by necessity that not all ideas make it through to final design, and he or she must be prepared to let go of ideas rejected with good design rationale. A recent study shows the latter case as more significant. In Thompson et al.'s (2005) study, when customers were given a choice of three models of a digital device of varying complexity (simple, mediocre, and highly complex), more than 60% chose the one with the most features. But when they were asked to use the digital device, *feature fatigue* set in, and they became frustrated with the functions they had. Thompson et al. claimed that we, as buyers, seem to overestimate how often we will use the overloaded functions, and that we also overestimate how easily we, as users, will figure out how to use these functions in the future, believing that designers would not make the functions difficult to use—an erroneous assumption. Likewise, Ouden's (2006) study also purported this *feature fatigue* that comes from poor usability, by which users imagine that the product does more than it really does.

An important lesson learned here is that a mobile product that does not have enough and interesting features may fail to attract customers' eyes in the store, but also too many does not guarantee the success of the product. In answer to the question, there is no easy solution, but perhaps we need to notice that a plurality of commercial companies (Apple™, Samsung™, Motorola™, Nokia™, LG™, Sony-Ericsson™ and so forth) have already figured out the crucial point that would make the complex things simple (such as Apple™ iPod). We seek the answer to how to do so in this book.

CLUMSY SCREEN SIZE – THE PARADOX OF USER NEEDS

The small screen size that limits the information that can be seen at one time is typically considered the restriction of the creation of a usable mobile interface, as Ziefle et al. (2007) empirically found. The restricted screen space allows only little information to be displayed at one time, and as a consequence this increases cognitive workloads, particularly for older users whose cognitive abilities (verbal[4] and spatial[5] memory) are less pronounced.

You might simply think, then, that designers should present users with a larger screen, but this is another self-contradiction. While most users are barely willing to read a long document on a small screen, they want to have a smaller and slimmer mobile device[6]. It is not empirically possible to determine when or why a user would be willing to justify the small screen size in preference to the small hand-held device. Hence, a potential misunderstanding that a larger mobile screen may resolve every mobile user interface issue in a single shot causes the frequent failure of creating useful and enjoyable experiences that mobile devices can uniquely provide: *portability* and the *carry principle*.

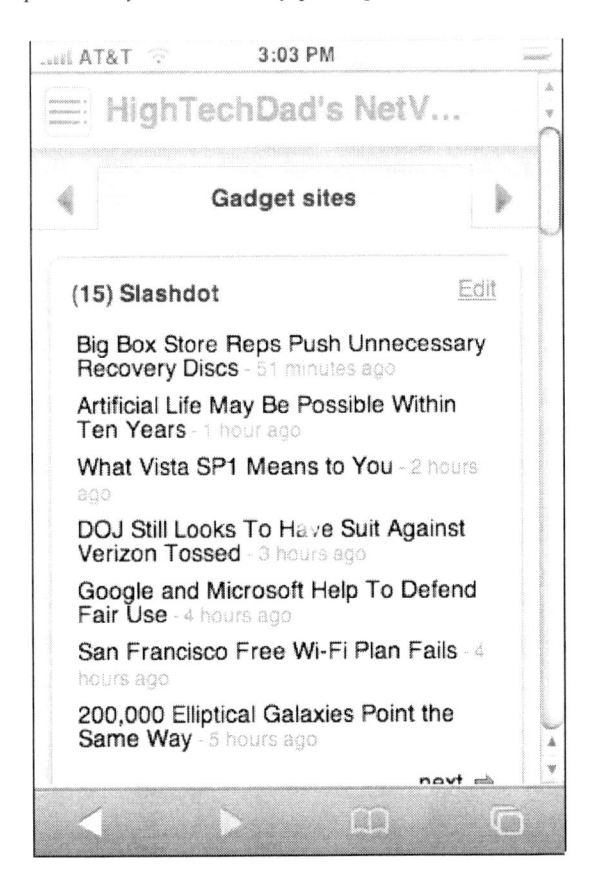

Figure 2.2. The tab-based interaction style in Apple™ iPhone. (Downloaded from techdad.net.)

---

[4] Ability with language, based on the premise that there is a deep relationship between intelligence and various linguistic skills
[5] This perceptual and cognitive ability enable one to deal with visual-spatial tasks.
[6] See AvantGo™ Mobile Lifestyle Survey (2004)

As a practical approach to overcome this small screen size, for instance, Apple™ iPhone adopted two technical advances: *tab-based interaction* style and *touch-sensitive* interface. The former allows the user to easily navigate through the tabs on the top (see Figure 2.2) or a handy navigation mechanism to show many Web pages opened, as shown in Figure 2.3. It certainly helps the user to locate what they are looking for and allows direct access to the target Web page. The latter effectively enlarges the screen, since the conventional physical numeric keypad is off the device, and replaces it with touch-sensitive input.

Still, it is barely sufficient for people to see much text-based information with capabilities such as Web page browsing. To address this visibility issue, several mobile handsets adopt a *full-browsing* presentation—that is, a full-size window first (as shown in Figure 2.3), and then if the user wants to focus on a specific region, he or she can enlarge the region. *Full-browsing* shows original Web pages on a mobile device's screen, so it is evident that people would have lower cognitive workloads with few scrolls and page shifts and the same visuo-spatial perception of Web pages as if displayed on desktop computer screen.

A variety of other techniques have academically been proposed for rendering as much information effectively as that which the desktop computer can hold. For instance, to effectively support Web page delivery in the mobile device, the *automatic Web page adaptation* technique holds the greatest academic interest. For the Web page—originally designed for a desktop computer—to be viewed on a mobile device, the font size and size of the images, texts, and hyperlinks need to be rearranged into an appropriate style (e.g., the original landscape layout of the Web page must be reformatted into a portrait layout in relation to the conventional specification of mobile phones) to fit into the mobile device's small screen. A very conventional approach to providing the same Web page content for different handheld devices is to prepare specific versions (formats) for different mobile devices, for instance, one HTML version for the desktop computer and its corresponding wireless markup language (WML) for the hand-held device. This approach is very straightforward, but probably highly labour-intensive and inflexible. Even worse, any changes in the content may result in knock-on effects in every related version, which is highly inflexible and may easily cause inconsistency between the HTML and WML page. The necessity of Web page adaptation is thus inevitable.

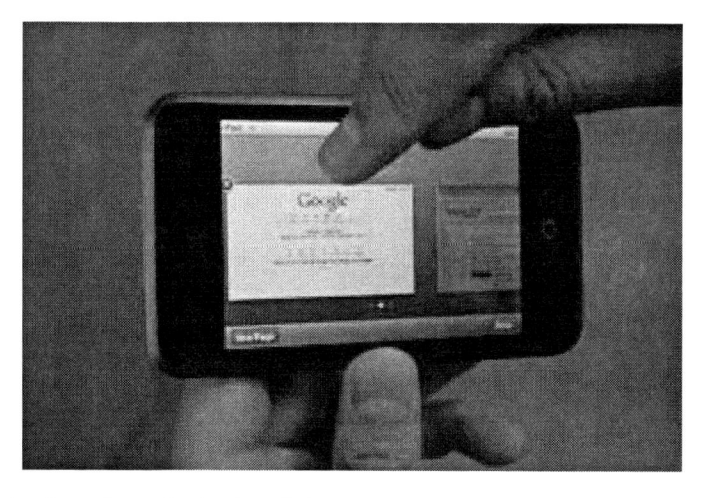

Figure 2.3. A navigation scheme in Apple™ iPhone.

To address this issue, Yang and Shao (2006) have proposed five Web page adaptation techniques: resizing, column-wise, thumbnail, replacing and transcoding. *Resizing* is a technique to reduce the content's shape to fit within the small screen size of the portable devices. *Column-wise* is a technique to transform content's presentation layout from multiple columns to a single column, which allows mobile users to view content from top to bottom as far as possible by reducing the movement of a horizontal scroll bar. *Thumbnail* is a technique to replace a large area of image with small icons to fit within a small screen size, while saving memory and transmission bandwidth. *Replacing* is a technique to replace rich media with text or voice in order to save money in relation to downloading the rich media or processing power to connect to the carrier network, and create an alternative when some types of media cannot be played on diverse platforms of handheld devices. Finally, *transcoding* is a technique to transform the rich media types with different modalities and fidelity to fit in with an individual portable device's computing needs. The first two, resizing and column-wise, adjust the original Web content by reducing the size of the images and text and turn a multiple-column layout into a single-column to provide better readability without losing information. In most cases, resizing and column-wise are the most primitive page adaptation solutions. Recently, thumbnail has become more widely adopted by many mobile phone user interfaces (see Figure 2.4).

Another approach that I have given much thought to as an enhancement is the combination of full browsing with a multi-view on a cubic hexahedron, as illustrated in Figure 2.5[7]. The interaction style is believed to allow the mobile user to change the viewpoint (or navigate Web pages) more easily using the full browsing screens projected on the four sides of the cube, so that any of the four information projections on the cube can be readily accessible with at most two fingertip movements. However, the limitation seems to be quite evident, with only four to six Web pages presented on the cube, but considering the current Web surfing pattern with the mobile device, this number would not jeopardize the usefulness of *multi-view interaction style*, virtually augmenting the limited small screen size.

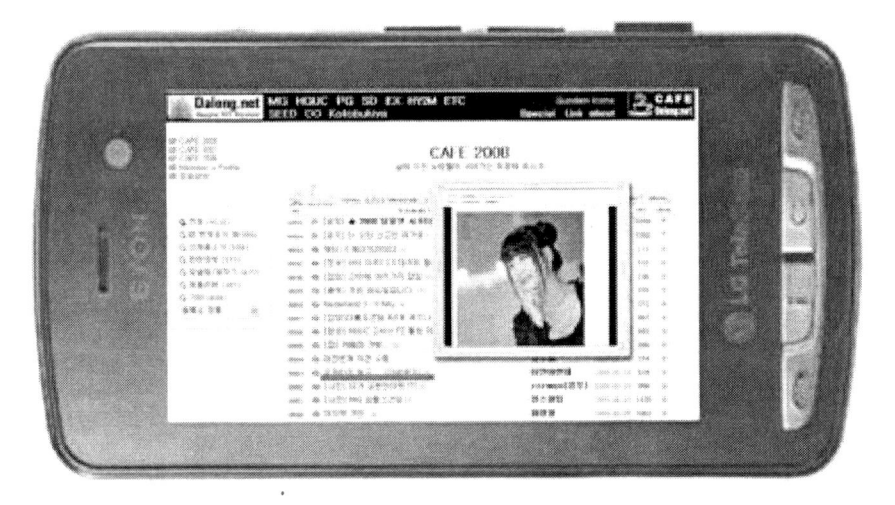

Figure 2.4. A thumbnail view in a mobile phone, downloaded from lge.com.

---

[7] It is quite similar to the virtual desktop environment in Linux™ or Windows™.

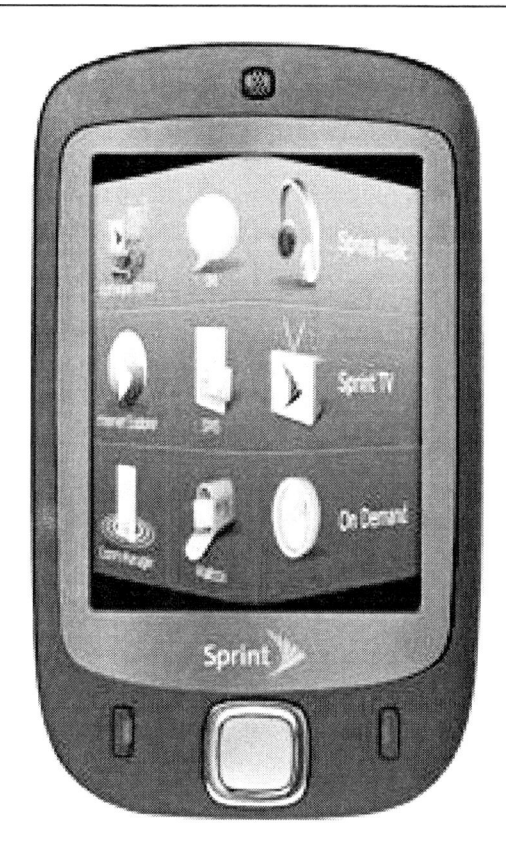

Figure 2.5. A multi-view interaction style in a mobile phone, Sprint's touch mobile phone. (Downloaded from Sprint.com.)

NO STANDARDIZED PLATFORM – DESIGN FREEDOM AND DEVICE PROLIFERATION

The desktop computing environment shares a number of common characteristics: all have a color screen of at least 800 × 600 pixels, a full-scale QWERTY keyboard, a mouse, and applications residing in windows. Connectivity may be generally speedier than that of mobile devices. Further, it is highly expected that the user of a desktop application is sitting on a chair or at least with a computer on the lap. There is a working surface, and both hands and cognitive attention are available to use the computer system. Interaction with other people takes place only through the computer, not generally in person around the computer. Compared to the common design considerations of the desktop computer environment, a relatively wide *design freedom*—which has been exploited by the designer community as immunity, permitting *designing whatever I like*—is a well-known nature of mobile user interface design.

Indeed, this *design freedom* in mobile interface design allows designers to enjoy less stress in making design decisions, but it does not mean that designers are less responsible for having developing further standard mobile interface design concepts. Ballard (2007) thus proposed the four basic concepts to designing a common mobile application to run on multiple devices, as follows:

- *Targeted* – selecting a set of targeted devices;
- *Common features* – selecting technologies and designs that will work on all devices;
- *Automated translation* – using a technology that automatically converts some standard core function (e.g., automatic Web page adaptation);
- *Class-based* – identifying groups of devices with common use and rendering characteristics, design the core function for each class separately.

Yet, *device proliferation* technically inspired by *design freedom* is a real challenge of mobile user interface design. Different devices have different capabilities and specifications: for instance, including virtual keypad, touch-sensitive screen, microphone, camera, motion detector, GPS (global positioning systems), infrared, Bluetooth™, Wi-Fi and so forth. And a different mobile carrier network does not provide exactly the same functions. An effective mediation of this *device proliferation* is thus a necessity in mobile user interface design, but it is not as easy as just saying so.

There are one or more mobile platforms[8] that can effectively manage this *device proliferation* issue, such as BREW™, WIPI™, Palm™, Windows Mobile™, Symbian™, Java™ 2 Micro Edition, messaging technologies (emails, SMS, MMS), and media environments (music and video, and camera).

Also, the common features of the mobile interface are worth noting here, which would help one to develop a more standardized mobile interface. These including the following:

- *Form* – devices are small and battery-powered, with some type of wireless connectivity, and small keypads and screens;
- *Features* – any information or entertainment features that might be desirable to have away from a computer or television itself, e.g., mobile TV (digital multimedia broadcast phone);
- *Capabilities* – the wireless connection, small size and power consumption have given devices slower connection speeds and slower processing;
- *User interface* – the relatively small screen, even though much bigger than before, limits the device to a single-window user interface, so sharing information between applications is problematic;
- *Fitness to the context* – a personal, always-present device needs to match a user's needs, desires, and personality reasonably well;
- *Always-available connection* – this has modified the experience of attending meetings and dinner with friends (e.g., refer to the "Blackberry Syndrome"[9]).

---

[8] A couple of years ago, Nokia could not sell its products in Korea since there was a regulation that forced Nokia to change its software platform (i.e., Symbian™) into WIPI, which was designed to protect local companies, such as Samsung™ and LG™. However, the Korean government is currently reviewing scrapping this regulation amid criticism that maintaining a fixed software standard would mean little when the global industry leans towards the adoption of open-source operating systems for wireless platforms (source from *The Financial Times* in July, 2008).

[9] The ability to read email that is received in real time, anywhere, has made the Blackberry™ devices infamously addictive. One of the market surveys reported that around 35% of Blackberry users would frequently check their mails even when they are dating out.

No Killer Applications – Going Mobile

While there are several factors that the successful mobile application shares with desktop applications, something they do not share is that they are not simply desktop applications exported to the mobile environment. A well-designed mobile function, when successful, is not a subset of the corresponding personal computer application, but rather an application whose features partially overlap and complement the corresponding PC application's features.

Let us muse on a mobile application now running in Bangladesh. In 2004, Bangladesh set up the *Rural Information Helpline*, based in the capital where Internet access is available, and a database of responses to common queries. Initially, however, many rural villagers were disconnected from the Helpline, though mobile phone networks cover more than 80% of the country. In response to this mobile paradigm, the *Mobile Ladies* initiative was introduced. These women, with a mobile phone in hand, go door-to-door in rural villages, listening to problems and advising on how best they can be solved, using the Rural Information Helpline via the mobile carrier network. This has been a huge success, showing that mobile devices can help connect the disconnected and address important social and economic needs. A key lesson to be learned is the ability of a mobile-based service to be truly user-driven, responding to current and future markets needs, in mobile applications design.

Indeed, unlike the substantial advances in mobile hardware, the context in which the mobile phone is used has not been significantly changed over the last decade, so very few applications have matched the success that SMS (short-text message service) has since seen. The rest of this section discusses three potential categories of further mobile applications, providing comprehensive coverage of mobile and wireless activity, related market opportunities in the near future, which will hint at challenges, and requirements for future mobile user interface design. (Please refer to juniperresearch.com for further detail.)

Mobile Learning

It is commonly said that *mobile learning* appears to be a promising approach to enable learning in context and analysis of real-world problems for the user of mobile devices and applications to better support their learning process, which will accordingly dictate the requirements of the mobile user interface. In the evolving educational landscape, mobile devices (e.g., mobile phones or Nintendo DS™[10]) appear to be an ideal solution to support education, since by definition that they are highly portable, personal, available anywhere and unobtrusive.

In this regard, mobile phones or game consoles were once banned as a distraction in classrooms, but now secondary schools, colleges and universities are exploiting mobiles as a learning platform. Those that double as Internet platforms, and iPods and MP3 players that can download video or audio files, mean students own a portable learning tool. What is more, the use of technology can be highly motivating, adding value and content in opening up new teaching scenarios. For instance, texting to and from students' mobiles is one of the major applications. A lecture theatre packed with several hundred students is an intimidating

---

[10] Cited in the *Guardian Weekly* 08 August, 2008.

environment for asking questions, so asking questions via texting resolves students' reservations about asking questions (Scornavacca, Huff, and Marshall, 2007).

Educational institutions have focused on delivering classroom-related information, such as lecture notes and review questions, only to personal computers. This is destined to change soon, since the market penetration of mobile phones grows rapidly and in some countries it is near or even more than 100%[11]. New technologies, such as cameras, media players, GPS, and Bluetooth™, are appearing in mass-market mobile phones, with many of these features existing even in entry-level phones. Also, the expansion of the 3G networks' coverage gives the opportunity to all the users of 3G-capable phones to have broadband access, anywhere and anytime.

However, not all kinds of learning activities and content are appropriate for mobile devices. Thus, interface design for mobile learning is necessity. In particular, I have long claimed that we should not simply apply known design requirements from desktop *e-learning* into the mobile learning context, because of its intrinsic limitations and benefits (Ryu and Parsons, 2008). For instance, in 1996, Jim Spohrer envisioned a portable educational device, the *Worldboard*, which would overlay the virtual and physical worlds (Spohrer, 1999). The Worldboard would give its user access to the history of his or her surroundings, make visible hidden features such as buried pipelines, and allow the user to add virtual annotations to that place. Such devices are now almost realized, as exemplified by the research of the Ordnance Survey in the UK, Augmented Reality and Zapper experiments (Brynat, 2003). As you can see in Figure 2.6, the interface design for this learning environment asks the designer to re-think a creative approach to render relevant information in a novel way.

Figure 2.6. Ordnance Survey Experiment with information overlays from mobile devices from direction sensitive GPS systems (Zapper). (Reprinted from Brynat [2003].)

Apart from the pedagogical benefits, mobile technology offers other benefits to mobile learning. For instance, the university can harness quality education practices by sending students information regarding timetables, exam dates and deadline reminders for

---

[11] See http://www.theregister.co.uk/2007/10/26/mobile_penetration_research

coursework. In addition, while on campus, students could use Wi-Fi to get information on library hours, renew books from the library, request books, and even browse the library catalogue from the comfort of a student café. In so doing, the interface design needs to take into account the *context* under which the user would work, as opposed to the traditional desktop-based user interface design, most of which are relatively context-independent.

There remain unresolved quite a few user interface issues in this area, which have to be addressed for mobile learning to present a beneficial and enjoyable learning experience, such as a practical screen size, effective user navigation scheme, tools for communication and information sharing between the learners, and benefits that the mobility offers in the learning process.

Mobile Healthcare

Healthcare is another domain in which great advances are beginning to be made in relation to mobile devices. It is not surprising that the healthcare industry has rapidly and enthusiastically adopted mobile technology. Healthcare professionals are routinely and increasingly using mobile devices to access and review patient records and medical test results, enter diagnosis and billing information for patient visits, consult drug formularies and other reference material, and synchronize information with their organization's centralized systems, all without the need for wired network connections (see Figure 2.7).

One of the successful examples of its kind was a *mobile data capture system* to enter health records by midwives, designed by the Apple™ Research laboratory in India. In this project, the Apple™ design team noted that each midwife was responsible for 3,500 to 9,000 people, whom they visited on foot or by bicycle, performing treatment of minor injuries and ailments, referring people to the local hospital, and providing ante- and post-natal care.

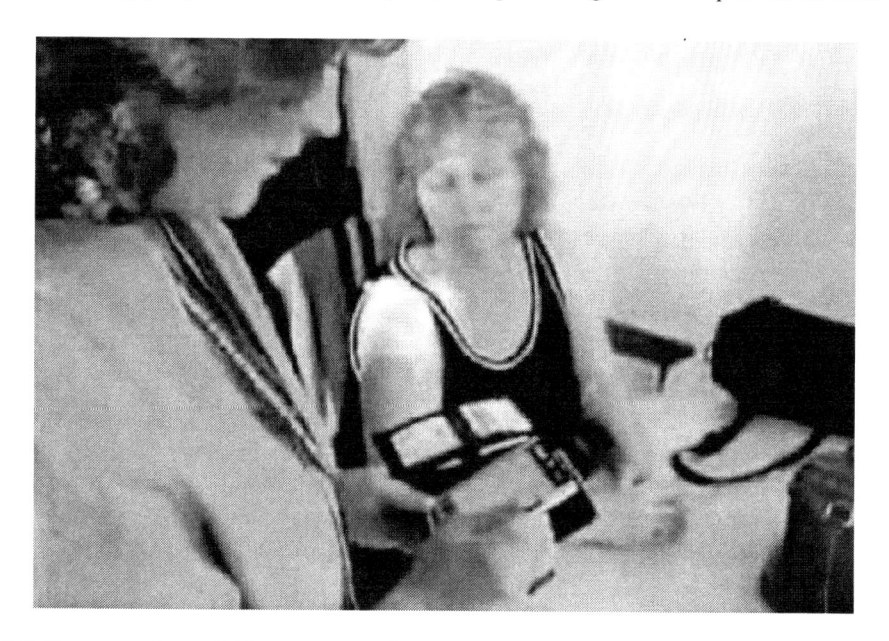

Figure 2.7. Data entry by a visiting nurse on a PDA.

In addition, at the same time, they had to keep a record of their daily activities and then compile them into a weekly and monthly report. By the observation of the Apple researchers, the midwives completed forms rapidly and recorded figures by hand; as a consequence, the forms were often incorrect, inconsistent or incomplete. The Apple™ design team set out the following design goals with the new mobile product:

- To translate the paper-based record keeping into an electronic format in a way that would fit into the working life of a midwife;
- To provide a navigational structure that would feel natural for the midwives; and,
- To provide a lightweight method for entering data.

A non-sophisticated technology can also "digitally-provide" in the health-related application. *Indian Blood Donors*, which operates throughout the country free of charge, links patients who need blood with those willing to donate it locally. SMS (short-message service)-based "bloodline" boosts this process. When someone needs blood, they send a text message to the "bloodline", and within seconds they will get a donor's name, blood group and contact details. The process then triggers an SMS to the donor with the contact details of the patient, and the donor and the recipient organize the blood donation between themselves. If the donor for whatever reason is unable to help, the parent or their family continue sending SMS messages to other donors on the list.

There remain unresolved quite a few user interface issues such as effective input, tools for communication and information sharing between mobile caregivers and professional health providers, and advantages that the connectivity offers in stakeholders' interests.

Mobile Business

For many people across the developing world, storing or sending small sums of money is economically impractical. This is due to the high cost and inaccessibility of banks and formal financial services. When all else has failed, several telecommunications providers, banks, and other companies have begun offering a variety of financial services via a basic mobile phone handset, e.g., WIZZIT in South Africa, Globe in the Philippines, and M-PESA in Kenya.

These m-banking systems support transfers of actual currencies. This means a person can walk into a mobile banking location, "cash-in" as if he or she were buying airtime for a pre-paid mobile account, and then transfer that money any time – often via text message – to merchants, utility providers or other individuals. Mobile banking reduces the need to carry cash, or to travel or wait in line to pay bills. It can guard against theft, replace costly bank cheques and increase the speed and reliability of transactions. However, each has a different set of actors and services, which implies the challenge of different user interface requirements. For example, some countries' laws strictly require stored value accounts to be managed by a registered bank, which requires a bank partner; in other cases, no bank is involved. The elegance of transactions via mobile headsets and text messages hides the financial services' complex organizational and technical capabilities, by which the simplicity and affordability make mobile banking a valuable service for poor people.

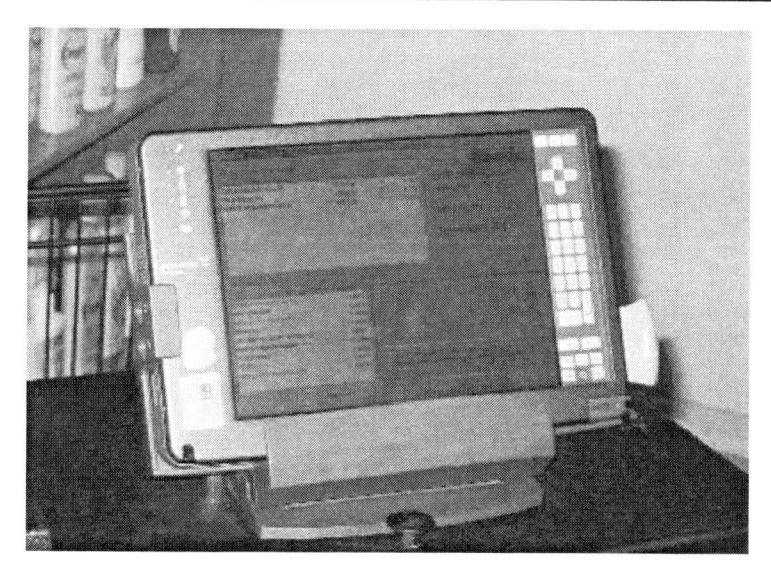

Figure 2.8. MyGROCER. (Reprinted from Gryazin et al. [2003].)

Another place for the *pick-up-and-use* mobile technology is the shopping centre. The MyGROCER system (Gryazin, Tuominen, and Kourouthanassis, 2003), for example, shown in Figure 2.8, uses a Tablet PC-type device built into a shopping cart, helping the user navigate to grocery promotions and to keep track of their purchases. Likewise, less sophisticated versions of this approach are common in supermarkets around the world. For instance, Tesco™, one of the biggest British supermarkets, launched its "Pocket Shopper" service which enables customers of its online store, Tesco.com, to make purchases from a menu of around 20,000 grocery products using handheld devices that run on Microsoft™'s Pocket PC operating systems. There are other technologies forthcoming that will further enhance these opportunities to reach out to consumers and help their shopping. The advanced mobile phones from Europe and Asia that store credit card data on handsets and read RFID tags are around the corner, too.

As a concluding remark, the key task that stands out in mobile user interface design is to figure out the differences that drive intrinsically different user interface processes. To do so, this chapter outlined the four challenges in mobile user interface design and, accordingly, how they define a somewhat different mobile user interface, though there is more than one vehicle for that explanation. The lesson to be learned is that the other design exercises cannot explain mobile user interface design itself, so the idea that we cannot put new wine into old bottles still holds true. The next chapter thus introduces the new bottles in mobile user interface design.

Chapter 3

MOBILE UI DESIGN PROCESS

It is no surprise that mobile user interface designers, having a vast number of key design decisions to make (the small screen, for instance, contains many individual items, each of which requires design decisions about layout, use of color, grouping, labelling of features and so on), are called the "Conductor of Information Design." They are not merely pioneering information architects themselves; their impact is even more far-reaching, as proclaimed by Herbert Simon: "Everyone designs who devises courses of action aimed at changing existing situations into preferred one." They inspire the most distinguished technology use in the innovative technological transformation.

In so doing, *design guidelines* can drive good design decisions. For instance, in desktop-computer user interface design, one can simply employ several user interface design guidelines to make their design decision, such as *Microsoft User Interface Design Guideline* (Microsoft, 1995), *Macintosh User Interface Design Guideline* (Apple computer, 1992), and *Web Design Guideline* (refer to www.w3.org). However, they only describe the general level of human-computer interaction design, which may be difficult to apply for appropriate mobile design decisions. Please note that the motivation of this book is to design new mobile interfaces, where there is a lack of well-established design guidelines; further, we cannot find general design guidelines to cover every technology in the world. If not at all possible with such guidelines, a *user-centred design process* will successfully lead designers, reviewing their design decisions through prototypes along with other members of the design team and potential users, which is the sole justification for a systematic design exercise as of now.

Having identified that the UCD process is the uniquely available systematic design process, in the book entitled *The Psychology of Everyday Things* (POET), Norman (1988) gives classic advice on how to design products that makes sense to users. To help every designer, he sets out four principles of good practice that promote the ease-of-use essentials:

- Ensure a high degree of *visibility* – allow the user to work out the current state of the system and the range of actions possible.
- Provide *feedback* – give continuous, clear information about the results of actions.
- Present a *good conceptual (or mental) model* – allow the user to build up a true picture of the way the system holds together, the relationships between its different parts and how to move from one state to the next.
- Offer *good mappings* – aim for clear, natural relationships between actions the user performs and the results they achieve.

Building upon Norman's design principles, other HCI researchers have also proposed user-centered design principles for effective interface design, such as Shneiderman's *Eight Golden Rules of Interface Design* (Shneiderman and Plaisant, 2005):

- Strive for *consistency* – make sure the system is internally and externally consistent, and be consistent with user's wider life knowledge and experience. This rule is the most frequently violated one, but following it can be tricky because there are many forms of consistency. Consistent sequences of actions should be required in similar situations; identical terminology should be used in prompts, menus, and help screens; and consistent color layout, location, fonts, and so on should be employed throughout.
- Enable frequent users to use *short-cuts* – as the frequency of use increases, so do the user's desires to reduce the number of interactions and to increase the pace of interaction. Mnemonics and special keys are appreciated by frequent knowledgeable users. Short response time and fast display rates are other attractions for frequent users.
- Offer *informative feedback* – for every user action, there should be prompt system feedback. For frequent and minor actions, the response can be modest, whereas for infrequent and major actions, the response should be more substantial.
- Design dialogs to yield *closure* – give users a sense of beginning, middle, and end in their interactions. Text messaging is a good example of a well-structured and simple dialog.
- Strive to *prevent errors*, and help users to recover from them.
- Allow *undo* – make it easy for the user to reverse mistaken input or actions.
- Make users *feel they are in control* of a responsive system.
- Reduce *working memory load* – the limitation of human information processing in short-term memory requires that displays be kept simple, and multiple page displays be consolidated.

However, living up to these ideals when designing mobiles can be an unreachable challenge. How do you make a system's functions visible and give good feedback when the device has such a small display and when users cannot always give it their full attention? Though the understanding of these ideals is undoubtedly a valuable one, some of the dilemmas and difficulties of using them in the mobile design exercise are still imminent.

The lack of practicality of the design principles has been addressed in many user interface design processes that also have been successfully applied in many years of design exercises. For instance, Norman's *seven-staged design model*, which will be further discussed below, proves useful. Also, in the practical design community, *Contextual Design Framework* (Beyer and Holtzblatt, 1998) and *Usability Engineering Approach* (Mayhew, 1999) are appropriate in commercial product development. However, note again that these cases are taken mostly from applications design for the desktop computer environment, and therefore we are not able to guarantee their values in mobile user interface design.

Designing-in these desirable qualities is not impossible, but it is certainly not an easy job. In this section, I propose a practical mobile user interface design process that I have long been using both in my industrial and academic fields.

NORMAN'S INTERACTION MODEL

The first cyclic design process of interaction, Norman's (1988) seven-staged interaction model (goal generation, intention to act, sequence of actions, execution, perceiving, interpreting, and evaluation), shows that the interaction between a user and a computer requires both mental and physical activities, and their distinct scope is a simple solution to the relevant design process. Physical activity relates to the motor actions performed by the user, such as using the keypad or the fingertip touch, to provide inputs. Mental activity refers to the cognitive processes associated with using the mobile devices. This also includes the processes required to plan for the execution of system inputs and to interpret and evaluate system outputs, as shown in Figure 3.1.

None of the HCI models have made such a substantial contribution to the designer's food for thought, for all sorts of lurid models to a great extent follow the track path of Norman's interaction model. It makes a contribution by placing interaction stages in the context of cycles of action and evaluation. Furthermore, the seven-staged model leads naturally to identification of the *gulf of execution* (the mismatch between the user's intentions and the allowable actions, i.e., goal-to-action problem[12]) and the *gulf of evaluation* (the mismatch between the system's representation and the user's expectations, i.e., effect-to-goal problem[13]).

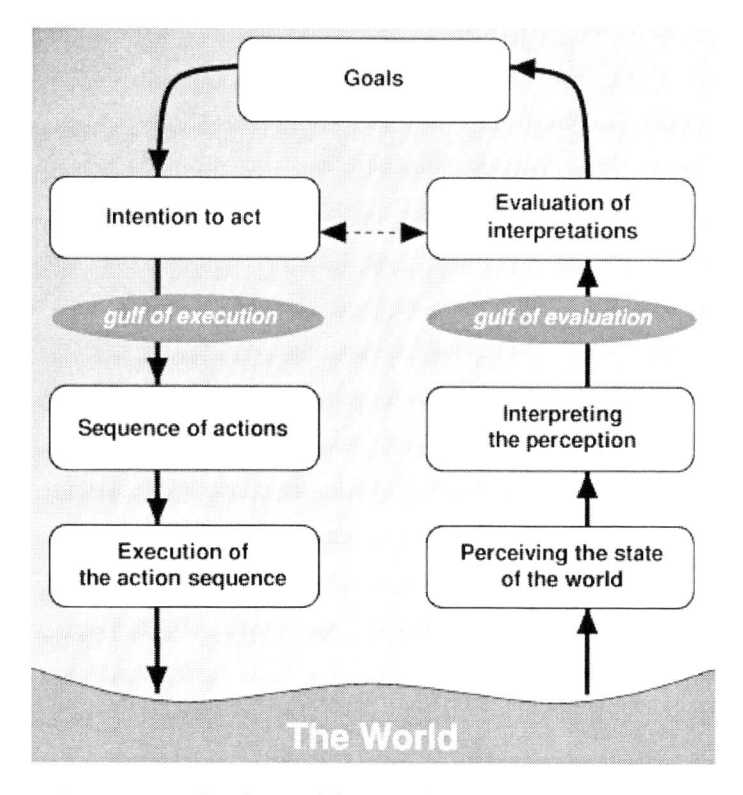

Figure 3.1. Norman's seven-staged action model. (Downloaded from w3.org.)

---

[12] See chapter 9 for further detail.
[13] See chapter 9 for further detail.

This model leads Norman to suggest the four principles of good interface design, and has evolved as the main theme in the user-centered design process, i.e., *design-evaluation cycle*.

In a nutshell, the design-evaluation cycle indicates that as designers consider how users will try to accomplish their goals, there are four critical points where user failures can occur: (1) users can form an inadequate goal; (2) users might not find the correct interface object because of incomprehensible labels and icons; (3) users may not know how to take a desired action, and (4) users may receive inappropriate or misleading feedback.

A Guide for Practitioners: Norman offers seven stages of action as a model of human-computer interaction. For instance, for a user to send a text message to one of his or her friends,

1. Forming the goal, *"Well...texting to Dave..."*
2. Forming the intention, *"Okay, using the mobile phone in my pocket..."*
3. Specifying the action, *"To send a text message... first I have to find the menu...and then... and then..."*
4. Executing the action, *"Click the big button on the keypad."*
5. Perceiving the system state, *"Oh... here it is..."*
6. Interpreting the system state, *"Okay. It's 'Message', and selectable with this big button..."*
7. Evaluating the outcome, *"Okay...now if I select this I can find other options to send a text message..."*

* This process continues until the completion of the overall goal – Texting to Dave

Though the seven-staged action model is a fine understanding for designers to take into consideration as they develop their user interface, it does not mean that practitioners are very willing to use it for their practical design exercises. Its lack of engineering process does not get their feet under the table to take this approach further. At least as far as a mobile user interface design process is concerned, Norman's interaction model has fallen considerably behind the more amenable design principles. Hence, the following section discusses how Norman's *cyclic interaction model* can be accommodated in a practical design process, which is supposed to be central to this chapter.

PUTATIVE MOBILE USER INTERFACE DESIGN PROCESS

Successful mobile interaction design presents many challenges. Interaction design is complex because it requires many different considerations that are difficult to make predictions about at an early stage of design, as proclaimed by Donald Norman in his book entitled *The Psychology of Everyday Things* (1988). This has driven the user interface (UI) specialist to undertake considerable effort (and research) regarding how a usable design can be obtained, in a piecemeal and practical way.

The practical design process has often been dealt with at several levels. For example, Rasmussen (1986) and Vicente (1999) have concentrated more on the job or task level, or the work environment of human-computer interaction, which is considered fundamental in presenting a new task model. In contrast, Card, Moran, and Newell's study (1983) focuses on system users and their activities, to cognitively describe users' interactions with a system.

Recently, *Activity Theory* (Nardi, 1997) has been proposed as a potential approach for identifying and relating the elements that should be taken into account when designing new work practices. However, the core of these design processes remains intact – the general design process is usually considered phased structure comprised of *design-evaluation cycles* with several iterations. This is well represented in Ballard's five-step mobile interaction design framework, as shown in Figure 3.2.

What one can notably see from Figure 3.2 is a mobile-phone–oriented development process, for instance, *emulator testing* (see Chapter 10 for further detail) and *pattern library* (refer to Chapter 8 for details) being widely adopted by many commercial mobile phone vendors. However, the first three stages remain almost the same as other user-centred design processes, such as Preece et al.'s (2007) interaction design process below. The first market analysis phase is similar to gaining understanding about the devices the intended user groups have and how the users select mobile phones in the competitive market, as heard from the intended user groups. In the next step, the requirements gathering process, use information about the targeted user group and their perception of the technical domain is collected to combine the user and device information with project needs, and organizational objectives. Likewise, Preece et al. (2007) define the process of interaction design with four basic *iterative activities* as follows:

- Identifying needs and establishing requirements for the user experience;
- Developing alternative designs that meet those requirements;
- Building interactive versions of the designs so that they can be communicated and assessed; and
- Evaluating what is being built throughout the process, and the user experience it offers.

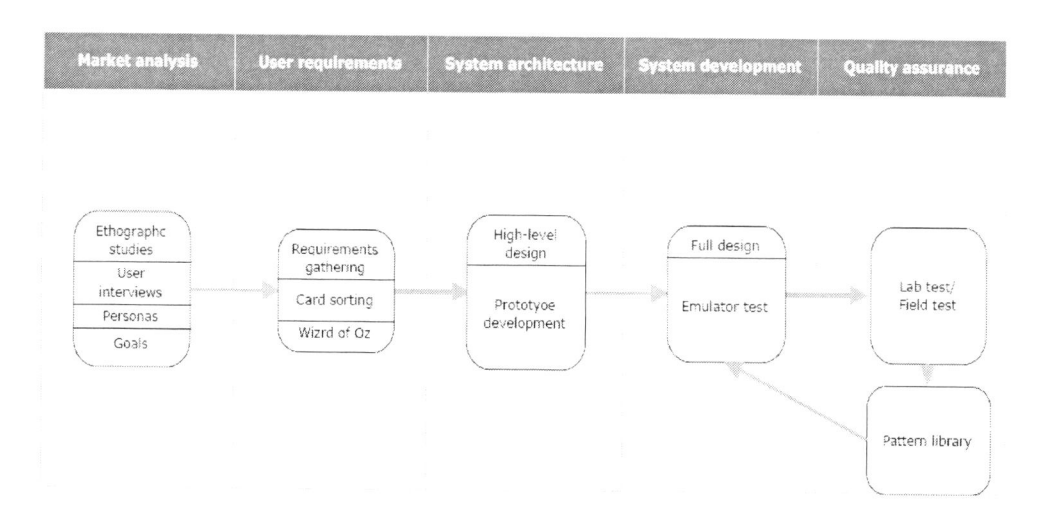

Figure 3.2. A mobile interaction design framework, extended from Ballard (2007).

Even though the words and expressions being used in both design processes are somewhat different from each other, the qualities and features to be considered are all but the same. In both design processes, three hallmarks of user-centred design should be noted. Firstly, note that these activities are intended to inform one another and to be repeated themselves until the final product is released, i.e., *design-evaluation cycle*. Secondly, they emphasize the findings from the different ways of engaging, and eliciting knowledge from users should be interpreted with respect to ongoing design activities. Finally, and most critically, evaluating what has been built is very much at the heart of mobile interface design. For example, for the final step, measuring the usability of what has been built in terms of whether it is easy to use provides food for thought that certain changes must be made or that certain requirements have not yet been met (Preece et al., 2007). In so doing, eliciting response from potential users about what they think and feel about what has been designed, in terms of its appeal, touch, engagement, and usefulness can help explicate the nature of the user interface.

Yet existing design processes seem to be impractical, as they are not able to clearly provide the mobile interface designer with the design rationales about what each stage must achieve—as well as what triggers that stage and with what techniques—which count most in the designer's favour. Instead, as illustrated in Figures 3.3 and 3.4, the *putative mobile user interface design process*, in combination with the refined designing procedures, seems to encompass hands-on design practices because the process can indicate appropriate design deliverables to the next stage and uncover relevant design issues in each design activity. Note that different stages may employ different methods or representations for their distinct deliverable (see Figure 3.4).

In detail, the putative design process starts by obtaining an understanding of the general context of the current work practices. Context of use relates to the *physical, social, cultural, technical* and *organisational* contexts in which the system will be used. The outcome of a context of use analysis provides the mobile interface design team with background information that can feed into the subsequent design and evaluation stages of the project being performed. In this stage, a description of the work environment relevant to mobile interaction design is obtained from the intended user population. *Interviews* and *Critical Incident Methods* (Flanagan, 1954; see more detail in Chapter 4) provide ways of gathering data on users' preferences and specific work constraints and so on, whereby the designer can observe how the user employs the technology in their own work environment (i.e., *ethnographical study*, which is also discussed in Chapter 4). As deliverables, user and environment profiles allow the designer to represent the current work practices in a more concrete style with some degree of abstraction. It also provides a flexible description for systematic reviewing of the work environment and its control structure.

In the second stage, an understanding of the work environment results in a description of current major tasks or tasks to be redesigned. Additionally, user and organisational requirements capture factors such as the important characteristics of the intended user group and their needs in relation to the system being developed. The result of this requirements analysis should be a clear description of the tasks that the system needs to support. In addition, the results of the requirements-gathering exercise should include a description of the functionality required to support these tasks. This stage should also provide the designer with a clear idea of the potentially new work practices in which the products or applications will be used. Here, several work analysis methods identify *critical tasks* in the current work environment. For instance, *hierarchical task analysis* (HTA; Annett, 2004, as a key reference) is a common form of *decomposition-based task analysis*. It generally uses a graphical representation of a decomposition of a task into subtasks and operations, or actions.

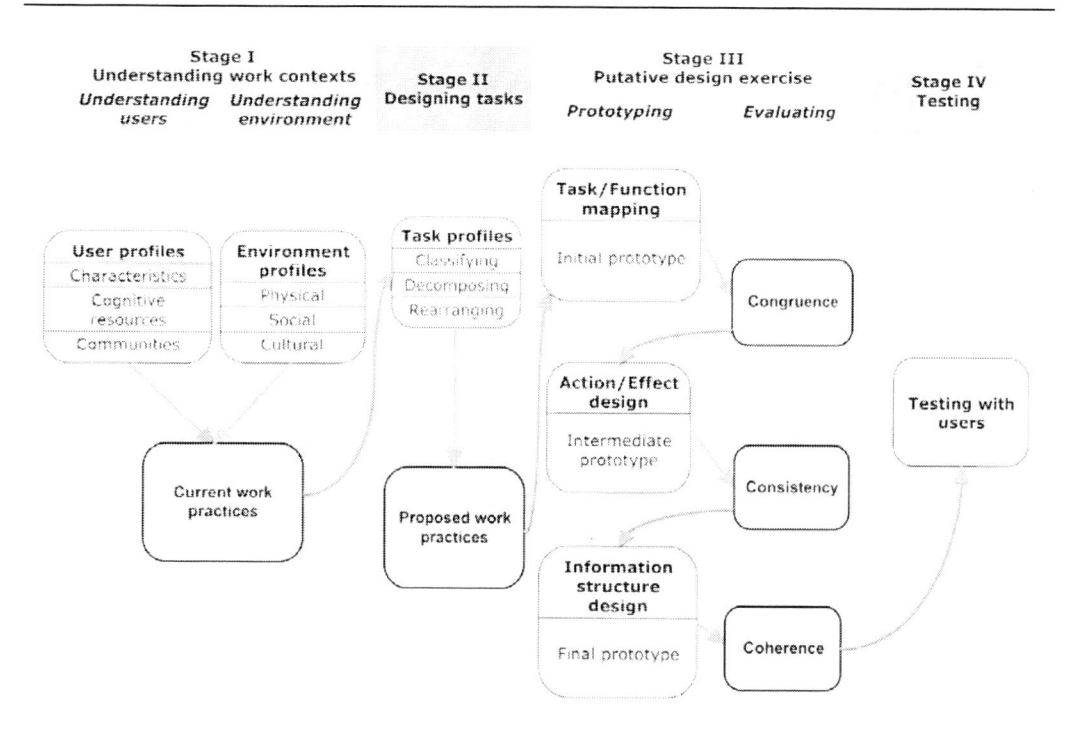

Figure 3.3. Putative mobile interface design process, focusing on what each stages achieves and what triggers the subsequent stages.

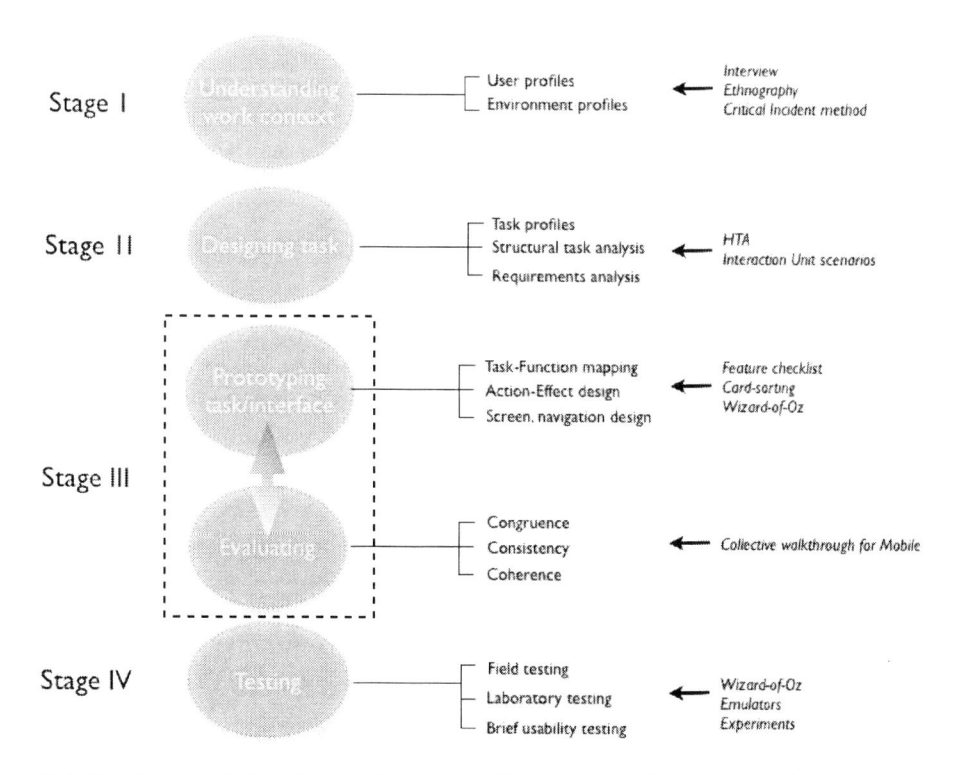

Figure 3.4. Putative mobile interface design process, focusing on techniques in each stage.

While HTA is concerned with presenting a description of the steps that are required in order to achieve a task, *cognitive task analysis* (CTA) elaborates the knowledge required to perform these actions. That knowledge may consist of the relationships among important concepts. *GOMS* (goals, operators, methods, and selection rules; Card, Moran, and Newell, 1983) is one of the most prominent models in the HCI field, indicating relevant design decisions to be made in terms of goal-action matching. Also, in my favour, *interaction unit scenarios* (Ryu, 2003a, 2003b, 2006; Ryu and Monk, 2004a, 2004b, 2005, in press) capture some representation of the requirements (either mental or physical, or both) that the user is supposed to have or that they need to have in order to achieve a task. Indeed, early HCI models (e.g., *TAG* [Payne and Green, 1986], and *Cognitive Walkthrough* [Lewis and Wharton, 1997; Polson, Lewis, Rieman, and Wharton, 1992]) have been to some extent successful in some aspects of task modelling. For example, TAG can be used if the designer is able to understand the task features that must be in the mind of the user. Cognitive walkthrough can reveal hidden issues behind a design, such as system effects that are insufficient to signal the completion of current goals. However, the assumptions made by the early HCI models are neither explicit nor usable for mobile interface designers with no full background in HCI. Moreover, they may not be able to deal adequately with additional cognitive issues that are inevitable in relation to mobile device uses. See more detail in Chapter 5.

Based on the first two stages, conceptual design of the major tasks (or proposed work practices) is generated in the form of a prototype. Here, one widely-noted feature is that designers create the prototype with their own "mental model" of a system. It is natural for them to develop their own prototypes, justified and preferred by themselves. Hence, giving up their own design ideas that have been invested their creativity is not so easy. However, a key quality of user-centred design is to take it through to final design (i.e., *design-evaluation cycle*), and designers must be prepared to let go of their own design ideas rejected with good design rationale. In this sense, the designer is the person who is least able to stand back from his or her design and assess how a first-time user would react to it. This is why we need some evaluation methods that are able to help the designers stand back from their design.

In the third "design-evaluation" stage, the prototype being developed by designers comes through *collective walkthrough for mobile* (see chapter 9 for further detail), by the designers themselves. Note that in contrast to other UI design processes, this stage fosters a concurrent approach, i.e., both prototype design and evaluation at the same time by the same designers. This is a premise of maximizing design performance; i.e., if the designers are designing a couple of tasks, and simultaneously evaluating them, then they would have insights on what to design next without making the same errors in another task design. This is why the putative design process puts these two otherwise separate steps together into one stage. Its benefits will be further discussed later. Also, the idea of evaluating *congruence, consistency* and *coherence* informs the putative design process from other interaction design frameworks, too. Although we suggest the use of heuristics or guidelines in the design decision if available in this stage, a main emphasis in this book is given to assess prototypes by *collective walkthrough for mobile*. This provides an opportunity to reduce the final design lead-time of novel interfaces rather than establishing its design guidelines in advance. This is central to our scope and purpose of the book: to propose a practical mobile interface design process.

Finally, the conceptual design arising from the iterative third stage is tested in the last stage and then evolves through testing techniques such as *formal* or *brief usability testing*, in which representative users attempt to perform representative tasks with minimal training and intervention. This stage and the third stage are conducted in iterative cycles until all major tasks in the final design are fully covered.

To be fair, the wide angle of the putative design process poses a special problem vastly extending the scope of this book, by which this book might not be a concrete guide to mobile user interface practitioners. Hence, instead, we give much attention to the second and the third stage, in particular, the *interaction unit* approach for both developing new task models and evaluating proposed designs against *congruence, consistency,* and *coherence.* This can also be justified by the fact that both work context analysis (i.e., Stage 1) and usability testing (i.e., Stage 4) are relatively well exploited in other texts such as *Contextual Design Framework* (Beyer and Holtzblatt, 1998) and *Usability Engineering Approach* (Mayhew, 1999). Being directed to the references above, the focus of this book is thus on both Stages 2 and 3, though this does not mean that we utterly overlook work context analysis and usability testing in this book.

A Guide for Practitioners: Putative Mobile Interface
Design Process, Methods and Techniques

- Stage 1: Understanding the work context
- *Methods: Focus groups, interviews, observation, and critical incident method*
- *Representations: User profiles and environment profiles documentation*
- Stage 2: Designing tasks
- *Methods: Focus groups, interviews, observation, and interaction unit scenarios*
- *Representations: Task profiles and structural task analysis documentation*
- Stage 3: Putative design exercises – prototyping and evaluating proposed designs
- *Methods: Collective walkthrough for mobile*
- *Representations: Task-function mapping, action-effect design, information design*
- Stage 4: Testing more detailed design
- *Methods: Usability laboratory, field testing, brief usability testing*
- *Representations: Paper prototypes, simulations, high-fidelity markup*

In effect, the putative design process for developing mobile user interfaces includes the following key features:

- A central focus on the people who will use the systems, on their preferences and requirements, considering the differences in mobile context;
- Building simple models of the users, tasks and technological systems;
- An amalgamating design process between prototyping and evaluation;
- And, an iterative design process.

Once again, it is thus worth noting that the core part of the putative design process is designing a couple of tasks first and then evaluating them, rather than a full design set of all tasks before advancing to evaluation and testing. This tactic is expected to provide us with some insights into how to design a better user interface by experiencing the evaluation perspective in advance, even though this exercise may require a great time effort at the very early stage of design.

In the personal communications with the Samsung™ mobile design team, I found that designing a new mobile user interface seemed to be much more complicated than in the past, thanks to their design intent to include no existing systems, and there are few references to exploit. Therefore, what they really want is a holistic picture to develop a user interface from start to finish—in particular, how to understand mobile users, their various capabilities (especially, cognitive abilities) and expectations, and how these can be harmoniously taken into consideration in the mobile system or application design. The next chapter discusses them.

Chapter 4

MOBILE CONTEXT ANALYSIS

In Chapter 3, we proposed that the putative mobile user interface design process involve four main "design-evaluation" cyclic activities, as follows:

- *Understanding work contexts* – having a sense of work context, people and environment to gain a rich picture of what makes up the details of their lives, the things they do and use. In this stage, firstly, we have to answer the question "*Who will use this system?*" That is, the mobile interface designer should set out what the users want to use the mobile device for – what work practices they want to currently perform when using the system. Secondly, in answer to the question "*What is important to know about the work that people who will use this system to do?*", say, what characteristics of the user could have a significant effect on their performance with the system, for example, age or their prior experience with other devices, or their different cognitive capabilities. Finally, another key aspect considered is an understanding of the environment in which the users are employing the mobile technology (a.k.a. context of use), as this can have a major impact on their ability to interact with the mobile device or application in an effective, efficient and satisfying way. *Physical*, *cognitive*, *cultural*, *technical*, *organizational*, and *social* environment should thus be described in here.
- *Designing tasks* – determining what tasks or works should be designed in the user interface, and sketching out the task procedure in the prototype. At least three types of critical tasks (most frequent, most onerous, and new tasks) should be detailed, as you determine the tasks (or work) in need by each target user group. Once the mobile interface designer has taken into consideration the needs of the user, the next stage is to prototype a system that meets the needs that have been identified.
- *Prototyping task/interface mapping and evaluating a proposed design against congruence, consistency and coherence* – representing a proposed interaction design in such a way that it can be demonstrated, altered and discussed, and each prototype is a stepping stone to the next, better, and refined design. Also at the same time, relevant evaluation techniques identify the strengths and weaknesses of a design. The crucial point in this step is to carry out the design and evaluation at the same time. It is important as part of the design to keep completeness in order to give designers a substantial insight to perform subsequent design exercises.

- *Testing* – testing whether the product being developed is usable by the intended user population to achieve the tasks for which it is designed. On the basis of this feedback the designer receives at this stage, an updated version of the interfaces or applications produced will probably be necessary.

To a greater extent, successful interface designs (e.g., arguably, Apple™ iPhone) often find designers who have developed remarkable insights into the way people actually work and learn. For instance, the touch-sensitive interaction style of Apple™ iPhone has long been perceived as one of the least likely interaction styles, thanks to its relatively slow and inaccurate reaction and poor haptic feedback in clicking buttons through physical fingertip movements, particularly demonstrated in several empirical studies (see Hinckley and Sinclair (1999) for further detail). However, the Apple design team has challenged this, overturning the academic criticisms on the touch-sensitive interaction style. Though I am not part of the Apple design team, my personal experiences in other commercial projects indicate that these insights most often come from a close connection between designers and users. Designers who spend more time with users—observing how they work, understanding who they are, and testing design concepts and prototypes with them—are most likely to prove successful in creating interfaces that are a delight to use.

Awareness of cultural differences is also an important concern in mobile interface design, particularly for mobile services intended for a diverse range of user groups from different ethnic backgrounds. Along with standard differences in the way cultures communicate and represent information, designers from different cultures often use different form factors, icons, images, and graphical elements when creating products and dialogue features for an interface (see Figure 4.1).

Figure 4.1. Different mobile phone keypads. The left one (downloaded from Samsungmobile.com) with the round-shaped "OK" button in the centre was loved by the Korean market; in contrast, the right one (downloaded from Nokia.com) with an oval-shaped confirmation button (with no special label on it) was quite successful in Europe.

Work context analysis for mobile user interface design involves gaining a sophisticated understanding of users' work practices, goals, and contexts in order to make better design decisions throughout the subsequent design activities. For instance, I have been asked to design a *mobile ordering service* for the two intended user groups, and the user profiles of the primary user category below are part of the work context analysis. In this chapter, likewise, we focus on how to collect the diverse user and context differences, and what techniques would best serve the data-gathering activities under the putative mobile user interface design process. Equally applicable are typical methods including *ethnographic research, questionnaires* and user *interviews*.

- Customers are the people who come to Grand Harbour Chinese restaurant in Auckland to enjoy special Chinese foods and drinks during the opening hour of the trading day.
- Sixty percent of the restaurant's customers are Chinese and the other are made up of both local New Zealanders and other non-Chinese (around 20% are Korean tourists.) Most of the customers can speak English, but some of the local people and the people from other countries (mostly Korean tourists) find it hard to understand the names of Chinese meals.
- Customers come to the restaurant to have both lunch and dinner, so the busiest hours of the restaurant are normally between 11:00 a.m. and 1:30 p.m. as well as between 6:00 p.m. and 9:00 p.m.
- Customers normally have a meal with their friends, family members, business partners, or colleagues. Elder customers and children often come together with their immediate family members or parents to have a meal at the restaurant.
- About 75% of the customers spend at least 45 minutes in the restaurant, and they like to take their time and enjoy their meal in the restaurant. It normally takes them about 5 to 10 minutes to choose and order their favourite meals; however, it takes more time if they come to restaurant during the busy hours, because the waiters or waitresses are extremely busy and trying to take orders from each table as fast as they can. Therefore, they sometimes have to wait at least 10 minutes until a waiter or waitress reaches their table and takes an order from them.
- All of the staff (except two managers and three chefs) are young people aged 18 to 23, and they work part-time. There are 20 waiters/waitresses in the restaurant; 11 of them can speak both Chinese and English, and the rest of them are New Zealand natives who speak only English.

While *focus groups* (see the section below for further detail) appear to be commonly used by interface designers, they often do not yield good results for informing design decisions (because they are mostly done outside the user's physical environment and away from his or her work). *Ethnography* would then be a good discipline. A typical ethnographical study involves observing potential user's entire context, patterns, practices, and needs associated with work contexts. A possible application of ethnography in mobile phone design would be investigating how people interact with and share music, and when they need to change ring tones, which enhances their experience and fits their needs. It is particularly beneficial for creating new products, as users cannot yet articulate their needs and context for a non-existing product. This will be further discussed later. The excerpts above were obtained through an ethnographical study and several user interviews in my two visits to the site.

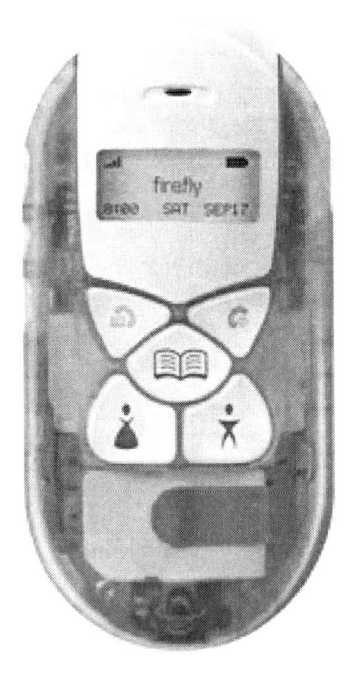

Figure 4.2. A phone for both kids and parents. The primary users are definitely kids, but the buyers would be their parents. These two primary user groups must be analyzed in this case. (Downloaded from fireflymobile.com.)

There are a variety of users' contexts to be collected in mobile interfaces design. Some mobile interface design will thus require several interviews under a variety of physical contexts and perhaps with diverse groups of users as well as individuals to capture some of the subtleties of mobile device use. In such cases, user *interviews* are a relatively inexpensive method of gathering some user data. For instance, by interviewing with both kids and parents, the Firefly™ phone (see Figure 4.2) a compact mobile phone for kids (as users) that also appealed to parents (as buyers) was realized. With just five keys, Firefly™ phones support ease-of-use by kids, and PIN (Personal Identification Number) protection allows parents to limit incoming and outgoing calls to numbers stored in the phone book (a kind of parental discretion).

> The primary user category (in the supermarket management system) consists of two user groups, which are "stocktaking" and "restocking" staff. Stocktaking staff are mainly responsible for checking products that are in shortage on shelves and recording the amounts that need to be restocked based on their personal experience. In comparison, by scanning the restocking task list provided by the stocktaking staff, the restocking staff restock the empty shelves with the relevant products. The status of the task in the list can be changed by the restocking staff, if necessary. For instance, when a restocking task is done, the restocking staff change its status to "completed" so that the task will be removed from the list by reloading the database next time.

The secondary user category (in the supermarket management system) is the managers of the supermarket. Users under this category access all of the functions provided by the current desktop computer system. They normally supervise their staff (the stock-taking and re-stocking staff) by monitoring the whole list of task activities.

For security and management purposes, each individual has his or her own unique identification number and password to log in to the computer system. Each type of user group in the supermarket has different privileges to access functions of the system.

The diversity of human abilities, personalities, and work roles or styles challenges mobile user interface design too. As illustrated in the excerpts above taken from one of my projects recently completed, it is not surprising to know that there exist many differences in users' roles and tasks with a system given. Thus, understanding the differences (e.g., the stock-taking, restocking staff, and managers in the example above) is vital and should be the most important input to our mobile user interface design process. Further, accommodating diverse human perceptual, cognitive, and motor abilities is also a challenge to every designer. For instance, some viewing angles and distances in a mobile interface would make the screen easier or harder to read. And the light reflection in the warehouse-like environment above would also lessen readability of the information on a mobile device. As a consequence, understanding physical and mental activities in which the target user group would like to employ the mobile device presents a solid cornerstone of a user-friendlier user interface.

Mobile work context analysis, in effect, is the process of learning about users by observing, interviewing, and surveying the actual users in action. Please note that it is different from asking questions in focus groups outside the users' physical environments and away from their work. It is also different from talking with expert users who may claim to speak for ordinary users but often unknowingly misinterpret them. They often do not know what really happens as the user operates the device in the physical environment. Hence, the first task in mobile user interface design would always be to understand the users, because users decide whether to use a mobile device—not designers or technically early adopters. The more we know about mobile users, the better we can design for them. In so doing, we can identify the features and functions that the user would like to use. For instance, by the work context analysis of the supermarket example above, the features and functions in need we have identified from the current work practices are listed below:

(1) Human errors often occur when writing down the names or serial numbers of the products by the stocktaking staff. The results of these errors include the wrong products being brought out and other staff members having to double-check to make sure the correct products are called from storage when it happens.
(2) The piece of paper that contains a stocktaking staff member's hours of labour could be lost before it is handed to the staff member who restocks the shelves or before he or she records the information in the computer system.
(3) Manually altering database records may cause incorrect and inconsistent information being changed.

The aim of this chapter is to muse on some users' individual characteristics that could have an impact on how they use mobile technology, and their attitude toward mobile applications, devices and services. This is an important area to consider, because many designers of mobile information and communications technology appear to have a generic user group in mind when they develop a product, without a thorough investigation of the appropriate user aspects.

PULLING IT ALL TOGETHER

Mobile devices are quite well perceived as *portable* and *personal* products. People always carry mobile devices, and the mobile device generally belongs to only one person, by which quite often mobile users would like to have customized user interfaces. For instance, ACCESS Platform™ supports flexible user interface themes that enable interface customization options, as shown in Figure 4.3. The importance that customization has for mobile phone users needs to be focused; certainly, the current commercial success of personalized ring tones suggests this will be the case.

Different users also navigate a structured information space in different ways, accessing only the parts they are interested in. In order to help the user navigating within the complicated information space using their mobile handsets, automatic categorization would be of value, as proposed by Kules et al. (2006). By necessity, both the customization and personalization of the mobile system require the creation of a user profile, which is a collection of attributes depicting the preferences of the user regarding various aspects of the user characteristics.

Figure 4.3. Customizing different mobile user experiences – ACCESS™ Platform. (Downloaded from Access-Company.com.)

In so doing, pulling all together subtleties of different mobile user's characteristics would allow the mobile interface designer to make better design decisions throughout the design activities.

Mobile User Characteristics and Contexts

Apart from the concerns of customization and personalization, there are many other reasons for mobile user interface designers to consider a variety of possible differences among their mobile users. This seems to be more important than the conventional desktop computer interface design, thanks to new potential activities that mobile technology would imply. For instance, the use of the mobile phone differs depending on users' ages. Recent research (e.g., Ling and Yttri, 2002) shows that the older users tend to focus on issues of safety and security. Middle-aged users focus on the coordination potentials of the system, like appointment calls and corporate emails. Contrary to these age groups, the younger users have the most distinct profile, using the mobile phone as an expressive medium for social purposes. Therefore, the younger users actually discover new meanings for the interface and explore new usage possibilities (Katz and Aakhus, 2002; Ling, 2004; Ling and Yttri, 2002).

Older users have a different set of user requirements from those of younger user groups, yet these needs do not appear to have been considered much by mobile interface designers, except in the context of designated mobile products such as the Easy5™[14] mobile phone. In relation to vision, if eyesight starts to deteriorate, it becomes increasingly difficult to read things on mobile devices. It is important in the mobile HCI community that effective research and design methods are available to help developers make products and services that meet the needs and requirements of older users. These methods should provide information on what specific functions they would like to see in a product or service (e.g., larger text on screen, larger buttons on handsets), why they would like to use a particular product, and what kinds of technological developments would improve their quality of life both inside and outside the home (Goodman, Brewster, and Gray, 2004).

Differences in cultural assumptions and behaviour all may affect how users interact with a mobile interface design. For instance, cultural divergences may make some basic metaphors used in a mobile interface design incomprehensible to a user with a different cultural background. As shown in Figure 4.4, an icon cannot convey a single meaning, so if this is not fully considered, it would be more confusing rather than helpful in different cultural contexts (Kim and Lee, 2005).

One notably different feature between mobile users and desktop computer users is that most mobile users are not sitting attentively at a desk, which means they are socially engaging with other activities, and moving. They may be in rush-hour traffic, in a meeting, in class, on a train, or walking down the street in an constantly interrupting. Thanks to the *mobility* (of course, many of us may not be mobile while actually using a mobile phone or device; however, it appears that we move between instances of using the mobile device), navigation through the physical world while managing obstacles and picking routes is a task that uses a majority of a person's *attention* resources. This implies that we thus may not ask for a highly task-focused (or onerous) interaction in mobile devices.

Equally important as being able to communicate is the social aspect that mobile technology can bring to mobile users. Mobile technologies not only allow connection with digital data, but they also contribute to forming new types of communities. This converse side is that mobile phone users are quickly available to remote friends, family, and clients. This, unexpectedly, has led to greater interruption by the "*networked-ness* among people," but it also enables people to feel more socially connected. Thus, an important characteristic of mobile users is that they are present and immediately available.

---

[14] http://www.silverphone.co.uk

REFERENTS	ABSTRACT SET 1	SEMI SET 2	CONCRETE SET 3
1. Call Log			
2. Message			
3. Downloads			
4. Voice Recording			
5. Web			

Figure 4.4. Three sets of mobile icon – abstract, semi-abstract, and concrete icon design. Depending on a user's cultural background, it is perceived differently. (Reprinted from Kim and Lee [2005].)

An online survey[15] showed that devices such as Blackberrys™ and Treos™ actually linked Americans more tightly to their work rather than liberating them, the latter being the original intention. Being constantly connected means more work for many. Turning off the phone or simply not answering it is one popular method for dealing with the phone's intrusion into life. However, an ethnographic research study has revealed that mobile users in Madrid think that it is socially rude to let a call go unanswered (Jones and Marsden, 2006). In many mobile phone use cases, being readily available means that people answer their phones, either with voice or text, in what used to be considered inappropriate places. *Texting* and even voice calls in a public toilet are becoming more common. Accepting a phone call during a personal conversation has become another common practice. Recently, a Korean carrier network launched a social connection service, called "*T-Friends*," which can a user's family, close friends or colleagues know whether or not he or she is interruptible. Likewise, *availability* allows the mobile device to communicate with instant messenger-like technologies (e.g., Yahoo™ Mobile[16]) with confidence that the user is present (or not present) and will receive the information immediately.

With all due respect, being mobile means that user location, physical and social context may change. Shifting context and navigation makes the mobile user more interruptible and easily distracted than desktop computer uses. This implies another important quality of mobile interface design—how to give appropriate contextual cues in such a hastily moving environment. For instance, a restaurant coupon application could send coupons at lunchtime when the user is away from home and near restaurants. Future mobile devices may have any

---

[15] Digital Life America survey in 2007.
[16] http://mobile.yahoo.com/messenger

number of other information sources that will be used to capture the current context of the user more accurately.

A few years back, most mobile phone users were very skeptical about Web surfing or watching television on their phone, since it would not be the same experience as afforded by the desktop computer or the conventional TV. However, an increasing number of Korean and Japanese mobile phone users are now interested in watching television on their phones, thanks to the new technological advance, so-called DMB (digital multimedia broadcasting). Around 23% of total mobile phone sales in Korea accounted for DMB-supporting phones[17]. As a typical frame of reference, a long commuting time in Seoul or Tokyo allows watching TV as an acceptable and entertaining option under only that physical and social context.

In effect, in mobile work context analysis, understanding what people do and think is essential. A main reason for having a better understanding of people in the context in which they live, work, and learn is that it can help designers understand how to design interactive products that will fit those requirements. Understanding the differences among people can also help designers appreciate that one size does not fit all. Learning more about people and what they do and think can also reveal incorrect assumptions that designers may have about particular user groups and what they need.

A Guide for Practitioners:
In mobile user interface design, user analysis may require the following information
(Jones and Marsden, 2006):

- What types of devices the user has;
- How often they are changed;
- What carrier and type of plan the user has;
- How the user makes decisions on which devices to buy;
- How many people are contacted using the device, and by which methods;
- Frequency of contacting these other people (to get a measure of how intimate the user is with the device and how strong a social network the user accesses with the device);
- How the user personalizes the device.

To emphasize the unique work context of a mobile device, consider again the *texting* example. The very first idea of a point-to-point short message service (or SMS) began to be discussed as part of the development of the *Global System for Mobile* (GSM) communication network in the mid-1980s, but it was not until the early 1990s that phone companies started to develop its commercial possibilities. At first, people thought that the written texts seen on mobile phone screens were so alien that few wanted to use it as a communication medium, saying that the keypad was not linguistically sensible. None took letter-frequency considerations into account when designing it. For example, key 7 on the mobile phone contains four letters, "pqrs'. It takes four key presses to access the letter "s", and yet "s" is one of the most frequently occurring letters in English. It is twice as easy to input "q," which is one of the least frequently occurring letters. In spite of its negative impression in the beginning, the use of SMS blossomed in teen life in the late 1990s (Katz and Aakhus, 2002; Ling, 2004; Ling and Yttri, 2002). Its limited length of text and its cheap cost had guaranteed the popular use of this service. Instead of delivering a certain predetermined type of

---

[17] According to the Korea mobile market survey (2007)

information, text messaging by teens had been significantly customized, and their contents came to deal with everyday life within a very limited and compact set of resources. Instead of talking on the phone, all teens want to share their views via text messaging. Kasesniemi and Rautiainen (2002) reported that:

> while some teens retain the most important messages on their cell phones, others have begun a movement to counter the perishable quality of text messages. Many teens copy their messages into calendars, diaries or special notebooks designed from collecting SMS messages. This practice of message collecting is an important part of text messaging culture. (ibid., p.178)

In this way, texting with a mobile phone significantly adds to teens' social communities, but through different mechanisms and with different benefits. At least one of the reasons that teens hooked into texting was a tactic of consolidating and shaping their own culture and community. And the success of SMS can be said to owe to the features of texting which well support the teens' social characteristics: *condensing*, *affordable* and *easy to communicate* without further interruption. It can be seen why MMS (multimedia messaging services) or 3G video calls has not replaced SMS, which has proved to be successful for teens. While mobiles are making at least some people less interactive with their immediate surroundings and less social with people nearby, like the effect seen with email, they are having a secondary effect. The *always-available communication* reduces the risk of going somewhere alone, either through safety concerns or through group coordination challenges. This added freedom is allowing at least teens more interaction with a wider variety of environments and people than they otherwise would have experienced. Please note that we, as mobile interface designers, are now asked to consider this social interaction effect, which we could not have imagined in the past.

A Guide for Practitioners: General User Characteristics to Be Considered—
This kind of information may be of some value in interpreting their needs and requirements.

- Age
- Gender
- Location of their working place
- Any disability or hardiness
- Current job titles and years in this job
- Responsibilities
- Previous job experiences
- Native language
- Educational background
- Previous experiences with other mobile devices

A Guide for Practitioners: Age, Gender and Income to Be Considered in
Mobile User Characteristics

- Age

Research in cognitive sciences has shown that there is a relationship between age and technology adaptation. In short, younger people are more receptive in adapting changes in the way they learn compared with older people (Bosma, Boxtel, Ponds, Houx, and Jolles, 2003). Technology assimilation among older people is slower compared to younger people (Morris and Venkatesh, 2000). It suggests the age of users as a design decision criterion.

- Gender

Attitudes and perceptiveness of technology are different between the sexes. While males perceive technology positively and easily, females think of it otherwise (McKinney, Wilson, Brooks, O'Leary-Kelly, and Hardagrve, 2008). Lack of interest and participation of women in technology may prompt a new approach to include their interests and design preferences in the new mobile user interface design.

- Income

There is a possible correlation between income and technology. Since technology is not a public good, owning a mobile device itself will consume some of the income. For instance, Apple™ iPod Touch or iPhone is being sold to mostly to consumers with higher-than-average income, and this would be the same case for PMP (personal media players). Technology utilization, especially new mobile devices, will require an individual to bear the expenses incurred. Thus, their income level could hinder their technology adoption.

Cognitive Psychology and Its Implications in Mobile UI Design

Understanding users is a key aspect in the design of mobile systems, devices and applications. Before we employ practical techniques to collect user data, quite often the designer can learn many logical design foundations from previous psychological understandings, particularly from an understanding of the cognitive and perceptual abilities that underpin human decision making.

The human ability to interpret sensory input rapidly and to initiate complex actions makes mobile or any human-computer interaction system possible. A classical information processing view is that the human organism consists of certain sensory systems (eyes, ears, skin, and so forth) which operate as receivers of external inputs; these receivers operate to encode these inputs. This encoded information is then passed on, via abstract information processing channels, to more central systems. Here, the limited channel capacity of human perception represents one of the bottlenecks in human information processing with mobile devices. Hence, this section discusses how early psychological understanding, in particular the discipline of *cognitive psychology*, would be strongly implicative of the design of a mobile user interface.

As shown in Figure 4.5, the *Model of Human Processor* (MHP: Card et al., 1983) is an early simplistic theory of human cognition. It aims to combine current psychological research knowledge with a simple approach that can readily be applicable to the solution of applied problems, such as systems design. Here, the mobile user interface designer should digest two important concepts: *attention* and *working memory*, which have significant implications for mobile systems design.

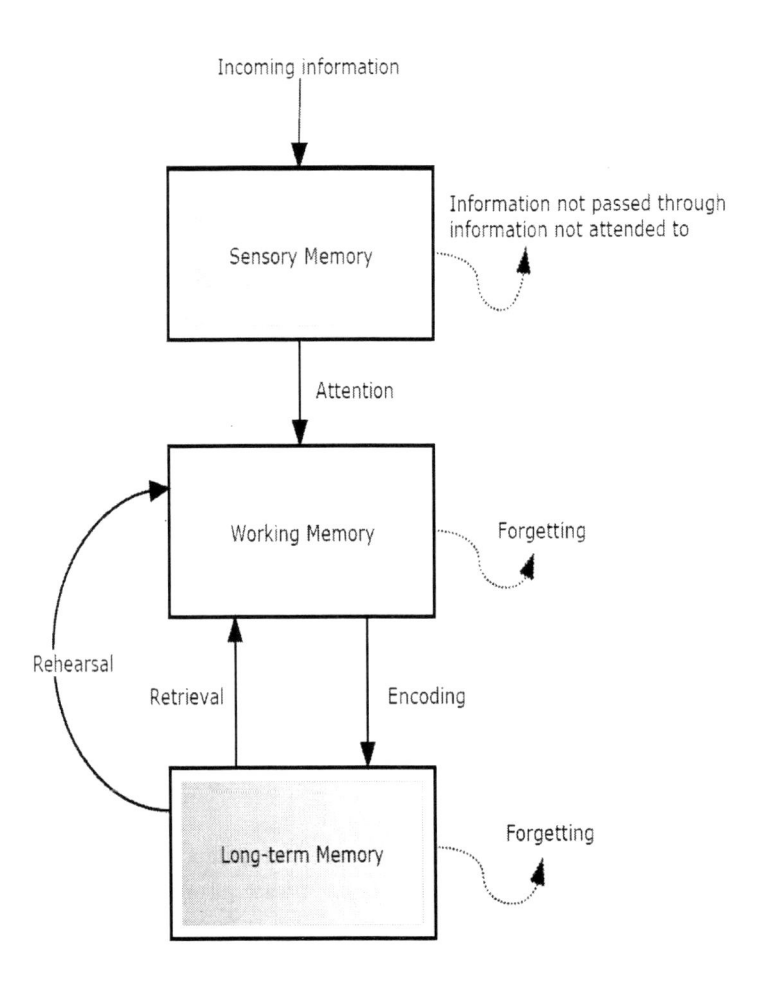

Figure 4.5. Model of Human Processor (MHP: Card et al., 1983), adapted from Newell (1990).

Visual Perception

Our information processing begins with *perception*. Our bodies have several sensors to detect sights, sounds, smells, and other physical contacts (e.g., haptic sense). Each information processing system creates a problem for higher-level cognition, which is to decide what to attend to from all the sensory information being received. Perception involves more than simply registering the information that arrives at our eyes or ears. A major issue concerns placing some sort of interpretations on the information being received.

Anderson explained the interpretation mechanism with Figure 4.6, "In reality, the figure is just a bunch of dots and ink blobs. ...However, many of you would suddenly see a dog in the picture, which means the dots and blobs are interpreted as a perceptively meaningful stimulus (information) on what you have seen." (Anderson, 2000, pp. 37-38). Simply put, what you see is not what is out there, but what is re-interpreted by your knowledge. As a consequence, it can be said that the information presented on the mobile interface would also be highly associated with this interpretation process, which includes two critical perception mechanisms: visual and auditory perception.

Figure 4.6. A sniffing dog. Reprinted from Anderson (2000, p.38).

Visual perception is the ability to interpret information from visible light reaching the eyes. The resulting perception is also known as eyesight, sight or vision. The details of this visual perception mechanism are far from the main objective of this book; however, object perception is closely associated with information design, particular for icons and menu design, which will be further discussed below.

For *object perception*, we tend to organize objects into units according to a set of principles called the *gestalt principles of organization*, after the gestalt psychologists who first proposed them. They claimed that our object perception is widely affected by where they are and by what surrounds them. Consider the various parts of Figure 4.7.

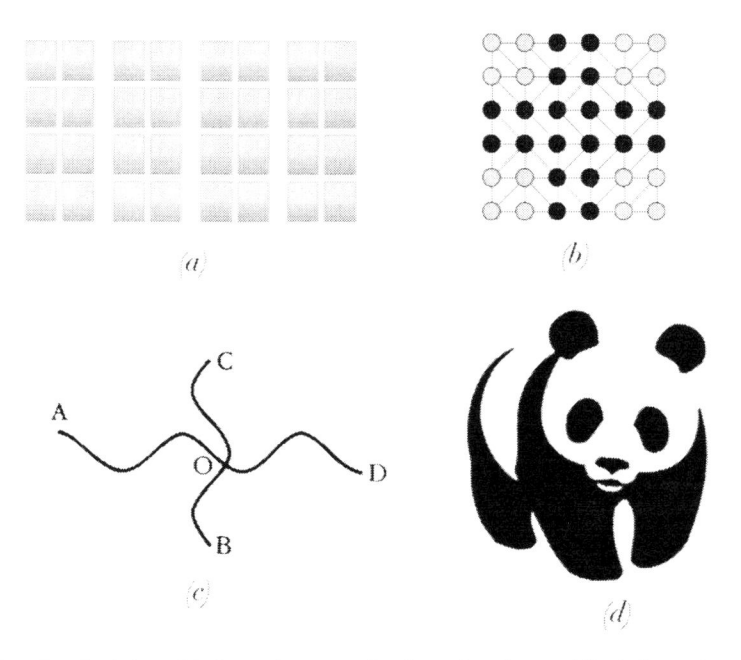

Figure 4.7. Gestalt principles of information organization related to mobile information design.

In Figure 4.7 (a), we quickly perceive four columns of blocks rather than eight separate columns. This picture illustrates the *principles of proximity*: elements close together tend to organize into units. Figure 4.7 (b) illustrates the *principle of similarity*, which means we tend to see these dots as the cross of black dots and the rest of white dots outside. That is, objects that look alike tend to be grouped together. Figure 4.7 (c) illustrates the *principle of good continuation*. We perceive two lines, one from A to D and the other from C to B, although there is no reason why this sketch could not represent another pair of lines, one from A to B and the other from C to D. However, the line from A to D displays better continuation than the line from A to B, which includes a sharp turn. Last, Figure 4.7 (d) characterizes the *principles of closure* and *good form*. We see the drawing as a panda, although each object could have many different forms.

Applying the gestalt principles to information design is more or less straightforward. Figure 4.8 shows one of the mobile interfaces employing the *principles of closure* and *good form*. The blank space makes the two sets of icons separate out. Further, it allows one to see the four icons at the bottom, i.e., "Music," "Videos," "Photos," and "iTunes" as similar, employing the *principle of proximity*, so the user can easily locate where to tap for playing media files without looking at the other icons. For more details of the visual information design examples, please refer to Tufte (1990) or the periodical *Information Visualization*[18].

Figure 4.8. The standing state of Apple™ iPod Touch, illustrating the *principles of closure* and *good form*. (Downloaded from Apple.com.)

---

[18] http://www.palgrave-journals.com/ivs

Auditory Perception

The auditory modality is different from the visual in two respects relevant to *attention*. First, the auditory sense can take input from any direction. Second, most auditory input is transient. A word or tone is heard and then it ends, in contrast to most visual input, which tends to be more continuously available and revisited.

Information design in mobile devices has overlooked auditory perception, except as warnings or alerting, in that we cannot close our ear, nor can we move our ear away. This is mostly due to the phenomenon of *visual dominance* (Posner, Nissen, and Klein, 1976) in our information processing with mobile devices. Notwithstanding, a loud sound will almost always get our attention, as it may signal a sudden environmental change that must be dealt with. As an amusing example, one of my friends chose an unpleasantly loud ring tone for his wife. As he hears the loud tone he readily answers the phone, which allows him to continue his focus uninterrupted, and so he will shift attention to it when heard, even if he does not pay much attention to the caller[19].

Yet, the sound input or feedback can annoy and startle, and their intensity can increase stress, leading to poor information processing in public spaces (Monk, Carroll, Parker, and Blythe, 2004). For sure, if you hire voice recognition technologies to command your mobile devices, it would be to a greater extent bizarre in a public space and vice versa in feedback design.

Attention and Information Design

When we are so involved in talking with someone else that we cannot attend to any incoming text message, *cognitive psychology* explains that our *attention* is selectively focused on talking with someone else. In such a case, likely important information (e.g., incoming text messages) may be ignored until the user attentively sees the mobile phone. The concept of attention explains how we are able to select information from our environment that we are interested in while ignoring other stimuli. Also, we often select inappropriate information, and then process it. For example, a mobile phone user sometimes selects the cues that stand out rather than useful or structural cues. Attention is the process by which the mind chooses from among the various stimuli that strike the senses at any given moment, allowing only some information to enter into consciousness, as shown in Figure 4.5.

Considering mobile user contexts, no easy task is attending to only one type of information source with the mobile device. For instance, while you are walking through a busy street, we need to pay attention to all the street information to avoid hitting other pedestrians. Thus the limits of our attention resources sometimes describe our limited ability to time-sharing performance of two or more concurrent tasks, and sometimes describe the limits in integrating multiple information sources on the mobile device. This is a hard question to answer what approach helps the mobile interface designer embrace the concerns of human attention resources, but by all means, the designer should bear in mind that mobile

---

[19] It is interesting to learn the "cocktail party effect". Imagine that we are standing in a crowded room while friends and acquaintances are socializing all around us. The sounds of conversations, glasses clinking, and music are loud and confusing. We are attempting to carry on a conversation in our little circle but are having trouble hearing the others speak. All of a sudden, from across the room, we hear our name mentioned. Immediately, selective attention operators spring into overdrive. We now find it easier to screen out other stimuli, to overhear it.

devices would be normally used whilst the user is on the go or at least not in a static position[20].

Occasionally, we are unable to concentrate on a single information source in the environment, in spite of our desire to do so; namely, we have a tendency to be distracted by another information source. A mobile phone user typing a text message while he or she is driving will encounter such a problem, a situation in which it is difficult to concentrate solely on composing a message. Another example is a user having difficulty locating the most relevant menu (or icon) in the midst of a busy display consisting of many blinking (or distracting) widgets, as exemplified in Figure 4.9.

It is not surprising that the user with the mobile interface would be not easy to locate the icon that he or she is looking for, selectively singling it out for their attention.

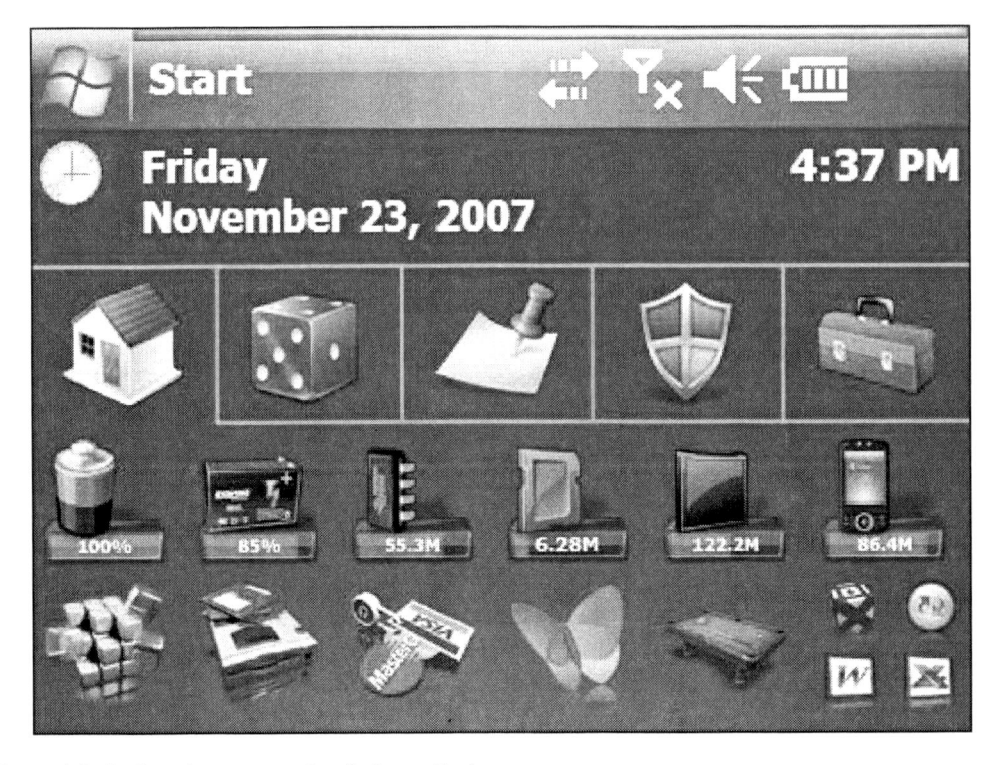

Figure 4.9. An imaginary example of a busy display.

Users know that they have to find and interpret relevant information as they perform a task on a mobile user interface. That is, our visual attention is driven by our need to attend, so *perception* and *awareness* is not quite passive and autonomous; instead, it is active and *goal-oriented*. In effect, attention is the set of cognitive activities that implement the selection of relevant perceptual input and the rejection of what is not relevant.

---

[20] For more detail, please refer to Dyson and Ryu's forthcoming book entitled *Cognitive Psychology and Its Implications in Information Design*.

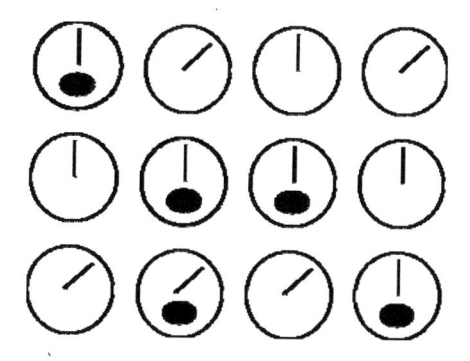

In the gauges illustrated above, for instance, we need to attend to two different sources of information at once—say, dials (represented by a needle) and lights (represented by a black dot). With the multiple information sources, we need to allow extra time and mental effort to interpret the multiple pieces of information at the same time, more than would be required by two separate sources of information. This account applies well to a *mode design* in mobile interfaces. Many mobile user interfaces employ modes because there are too few keys on the keyboard to represent all of the available commands. Therefore, the interpretation of the keys depends on the mode or state the system is in (e.g., in letter-entry mode, a click of the key "7pqrs" is highly expected to display a letter rather than "7".) Quite often, however, the user cannot notice what mode the system is in, because the mode signal must be perceived by more than one information source. Hence, many users find themselves triggering a sequence of commands quite different from what they had intended.

In the literature of cognitive psychology, five general conclusions of visual attention are described below. Building upon the summaries by Moray (1981, 1986), they have some implications for information design that are also of great value in mobile user interface design.

- *Understanding the user's mental model that guides visual attention.* Moray (1981) claimed that the user appears to form a mental model and uses it to guide visual sampling or perception. As expertise develops, the mental model becomes refined, and his or her visual sampling changes accordingly (e.g., a heavy texting user does not need to see the keypad in composing a message.) Therefore, the patterns of fixations would help the mobile interface designer arrange information on a display so that optimal performance results. For instance, frequently-used menu items should be placed centrally, and pairs of information that are often needed sequentially should be located close together. This would also be a benefit of *consistent interface* design, so that the user can readily establish a mental model to use the consistent interface.
- *Understanding the user's visual scanning pattern that guides effective information design.* To keep objects in *foveal* vision (the visual field perceives detail within about two degrees of the visual angle), the eyeball exhibits two different kinds of movement: *pursuit movement* and *saccadic movement*. The former occurs when the eye follows a cursor moving across the interface, such as following highlighted menu items with a navigation button on the mobile phone. Saccadic movements are discrete, jumping from one point to the next, such as searching an appropriate menu

item on mobile screen. For instance, in Figure 4.8, we can locate the "Music" icon by jumping between the icons rather than sequentially searching for it from left to right. The fixation in the saccadic movements is characterized by a location (the centre of the fixation), a useful field of view (diameter around the central location from which information is extracted), and a dwell time (how long the eye remains at that location). When the user reads texts on mobile screen, e.g., reading Web pages, both pursuit and saccadic movement are involved, Donk (1994) found that users were more likely to make horizontal scans than diagonal scans. This has many implications for information design, such as Web pages or multi-level menu structure. It has been said that we need to place the most frequently-accessed or important information at the top, so that the information can be easily accessed. Of course, since visual scanning is often internally driven by cognitive factors, there is no consistent pattern of display scanning and no optimal scan pattern in searching, beyond the fact that a search should be guided by the expectancy of target location. The Apple™ iPhone shown in Figure 4.8 helps this visual search based on expectancy, so the user can easily locate where to tap to play any media files without visually searching the upper part of the display. Also, certain display factors can attract our visual attention, such as items that are large, bright, colourful, and blinking. These salient items can be used to guide or direct visual attention. There is evidence also that search behaviour is sometimes guided by physical location in the display. It was found that when users exhibited a systematic scan pattern in searching for targets, they tended to start at the upper left. This fact may reflect eye movement in reading.

- *Human memory is imperfect, so design with a recognition-based interaction style.* Human memory is imperfect, and this indicates a prominent limitation of the mobile user interface design owing to the small screen size. Unlike the desktop interface, mobile user interfaces have many scrolling patterns or page shifts, which inevitably requires the mobile interface user to remember something to go ahead toward the next page. As discussed in Chapter 2, many technical supports can overcome this limitation, such as the *full-browsing* or *multi-view* interface, to minimize scrolling and page shifts.
- *Identify the useful field of view (UFOV).* To take in information detail, resolution of fine visual detail is important, which is an angle of no more than about two degrees surrounding the centre of fixation – the useful field of view. Several factors affect the UFOV. The size of information appears to be important. Also, aging tends to lead to a restricted UFOV. Scialfa et al. (1987) proposed that older adults take in smaller perceptual samples from the visual scene and scan the sample more slowly than young adults do[21]. When novice users perform a search in the mobile user interface, they examine each item in the set in a systematic order[22]. In the typical menu search task, they must locate a target word, symbol, or command, scanning the list until the item is located, and then press a key. Menus may be multi-level, in which case the target term may be reached only after a search through higher-level terms (like top,

---

[21] This certainly implies that the mobile user interface for the elderly should be differently designed, e.g., Easy5 mobile phone (refer to http://www.silverphone.co.uk)
[22] As we discussed above, this may not be true for some users, such as experts or the users who have very standard user interfaces.

second level and so forth). Mobile interface designers should thus structure a menu in such a way that target items are reached in the minimum time. Perhaps the items frequently used can be positioned toward the top of the menu as a consequence.
- *Use pre-attentive processing and Gestalt organization.* Many psychologists have argued that the visual processing of a multiple-element world has two main phases: a *pre-attentive phase* is carried out automatically and organizes the visual world into objects and groups of objects; then we selectively attend to certain objects of the pre-attentive array for further elaboration. Gestalt psychologists have made an effort to identify a number of basic principles that cause items to be pre-attentively grouped together on the display, e.g., *proximity, similarity, good continuation, closure* and *good form*. Displays constructed according to these principles have high redundancy. That is, knowledge of where one display item is located will allow an accurate guess of the location of other items in a way that is impossible with the less organized display. Because all items of an organized display must be processed together to reveal the organization, such parallel processing is sometimes called *global* or *holistic* processing, in contrast to the local processing of a single object within the display.

A Guide for Practitioners to Evaluate Information Design,
extended from Wickens (1999)

- Intensity – *use several levels, but with limited use of high intensity to draw attention.*
- Marking – *underline, enclose and so forth.*
- Size – *larger sizes attract more attention.*
- Blinking – *minimize the blinking animation in information design, unless it is a warning sign.*
- Color – *color contrast should be considered; for legibility, black text over yellow or orange background is better.*
- Audio – *use soft tones for regular positive feedback and harsh sounds for rare emergency conditions.*
- Pertinence – *a more highly likely event or information is given much attention.*

Working Memory

As depicted in Figure 4.5, the three memory (sensory, working, and long-term memory) stores differ in function, capacity and duration. Among them, the greatest interest by interface designers must rest on *working memory* (or short-term memory), with a greater emphasis on processing information when carrying out cognitive tasks. In particular, MHP (Model of Human Processor, refer to Figure 4.5) postulated a short-term memory with the following properties to model and predict human performance:

- *Function* – conscious processing of information where information is actively worked on;
- *Capacity* – limited (holds 7 ± 2 items);
- *Duration* – brief storage (about 30 seconds)

Also, MHP generates practical advice for interface design, including:

- Support *recognition* rather than *recall*, using graphics and labels.
- *Reduce working memory load*, presenting no more than 10 information chunks on a small screen.
- *Support the production of chunks* of items to facilitate memory storage, such as grouping of the icons "Music," "Photos," and "Videos," as shown in Figure 4.8.
- *Consistent interface design* would help relieve the user of the burden of fully remembering or activating the rules that should be recalled.
- *Training wheel* (Carroll and Carrithers, 1984) would also reduce the requirement for full employment of working memory by the novice.

Navigating through interface functions on mobile devices consumes cognitive resources. In most cases, you have to remember that, as people will be using these devices in dynamic social environments, they will be subjected to other factors that could distract them while they are trying to find information on mobile devices, including traffic noises, people walking past as they stand in the street, and so forth. Because these tasks require similar use of cognitive resources – *spatial working memory* (Baddeley, 1986) – they clash with each other.

Spatial working memory has been shown to be important for navigation in the real world, and also in an abstract information space such as Web surfing (Jones and Burnett, 2007). Even when alternating virtual and physical tasks quickly, either or both can suffer. This means that mobile users are more interruptible and spatially distracted than those in the desktop computing environment. It is defined that spatial-ability entails visual problems or tasks that require individuals to estimate, predict, or judge the relationships among figures or objects in different contexts. More specifically, it relates to individuals' abilities to search the visual field, comprehend forms, shapes, and positions of objects as visually perceived, form mental representations of those forms, shapes, and positions, and manipulate such representations mentally (Carroll, 1993).

This ability to recognize and handle spatial relations of objects has been significantly studied by a number of researchers in the area of HCI (e.g., Vicente, Hayes, and Williges, 1988; Vicente and Williges, 1988). For instance, Vicente et al. examined user performance when accessing files from within a hierarchical file structure, and found that high spatial ability users completed tasks more quickly than low spatial ability users. Stanney and Salvendy (1995) also found that users with low spatial ability found it difficult to construct a visual mental model of the system they were interacting with, thus leading to poor task performance. In effect, differences in spatial visualization ability led to certain users performing more efficiently than others at information search and information retrieval. The difference in task performance does not mean that users with low spatial ability cannot find information, but that they tend to be slower at doing so. Spatial ability is also not completely static; it can be improved with practice. However, since the onus in the design of mobile systems is on the designer to provide systems that can be used by the majority of users or customers, compensating for low spatial abilities in the target populations is generally considered to be a good idea, as shown by Morrison and Tverksy (2001). They demonstrated that spatial information can be improved when it is represented with graphics or when accompanied by animation. Interventions that help those with low spatial abilities on the mobile screen include spatial organizers like structure previews (see Figure 4.8—Apple™ iPod Touch shows a kind of menu structure preview), which can improve the performance of people with lower spatial ability while not hurting those with higher spatial visualization ability. The interface by reducing the number of hierarchies between menu items also improves the performance of low spatial visualization individuals while increasing the performance of high spatial visualization individuals to a slightly lesser degree.

Though previous findings are mostly made in the computer-mediated environment, spatial working memory has immediate implications for mobile interface design. Mobile devices have very small screens, and this fact has a knock-on effect on the amount of information that can be represented to the user at any one time. Consider menu hierarchies to let a user navigate different functions of a mobile device. Ziefle et al. (2007) demonstrated that a person with higher verbal and spatial memory capacity had a better performance rather than those with weaker abilities in this area. In particular, they (ibid.) showed that the elderly might have trouble using hypertext structure interconnected by links rather than using hierarchical menu structure, owing to their different cognitive capability. Hence, the menu structure would be accordingly designed for the intended user group. That is why mobile phones for kids or the elderly have fewer functions. Likewise, in relation to spatial working memory, Lee and Ryu (2007) empirically developed a map-based mobile interface for controlling a service robot, which proposed several design guidelines that might be of great value in developing a geographical mobile user interface design.

Other issues from the discipline of cognitive psychology, such as problem solving and knowledge representation, which cognitive psychologists have long considered important, are not further discussed in depth in this book. For more detail, please refer to Dyson and Ryu's forthcoming book entitled *Cognitive Psychology and Its Implications in Information Design*.

Mobile Environmental Characteristics: Physical, Social and Cultural

There is little doubt that mobile users would have different environmental characteristics from desktop computer users. One widely noted feature is that mobile devices are portable and personal belongings, so users can easily carry them, which indicates the fundamental difference between the mobile and desktop interface. Though *portability* gives users a certain level of work freedom, too often it makes interface design much harder, presenting us with substantial challenges as to how to design mobile interfaces to effectively work with surrounding work contexts, which has been little considered in desktop interfaces. That is, it is not easy to depict a common trajectory of the work environment with the mobile device. In this sense, when conducting mobile work context studies, one must examine the context of use that relates to the *physical, social, technical, organizational,* and *cultural* contexts in which the mobile system uniquely defines itself.

The aim of the *mobile environment analysis* here is to harbour the whole design process in a solid base of the mobile user data set. This background information then can feed into the following design processes, particularly for prototype design and evaluation stage, capturing user and contextual requirements such as important environmental considerations or what types of users would have significant problems in using the system proposed in relation to such contexts, and so forth. In effect, the mobile environmental profiles will serve the same function as user profiles: to enhance the lists of user environments and make them more detailed and memorable for the design team.

Physical Environment Characteristics

We are well aware of that people are influenced by their own activities, the activity around them, the physical characteristics of the workplace, the type of equipment they are using, and the work relationships they have with other people. If the products you design do not fit into the physical environment (e.g., a chopstick for eating soup), they may be difficult or frustrating to use. Hence, the physical aspects of the environment are likely to be immediately significant in mobile interface design. It is not surprising that the most elegant mobile design is compromised by the noisy outdoor environment, and that compromise will eventually lower performance, raises error rates, and discourages even motivated users.

A Guide for Practitioners:
When you perform the physical environment analysis, note that:

- Recognizing that the users' physical environments may substantially affect the success of your design.
- Looking for as many clues as you can observe or discuss with your users will help you avoid design mistakes.

For instance, in developing a PDA-based "Supermarket" job dispatching system (this example was discussed above and continues in this section), our design team summarized the physical environmental characteristics of the workplace as follows:

We have interviewed and observed the three groups of users: the stocktaking staff, the re-stocking staff and managers of the supermarket. Like other warehouses, the workplace in which our user groups are working is a huge indoor shopping department. There are rows of high shelves of products. In most places, lighting equipment is well located to supply a bright environment for the staff. However, there are some corners or places where staff may find it difficult to work because of poor illumination.

Here, the work of shelf restocking is completed overnight to provide a friendly and neat shopping environment for the next day. Obviously, at night, general illumination, which is commonly used, will not be sufficient. In particular, the readability of letters or characters (usually black to provide strong contrast) will be affected by the brightness of lighting.

It is hard to imagine that the current desktop computer system takes around 15 minutes, logging onto the system, opening a cascade of directories and finally getting to the object file to work with.

As staff members (especially, stocktaking and re-stocking staff) have to keep moving around to finish their daily job, some trade-offs have to be made to compromise required accuracy and the amount of information to be entered.

A Guide for Practitioners:
The following items are to be considered in the physical environment analysis:

- Light level – *intensity, illumination, contrasts, reflections, and quality of light*
- Noise level
- Size of the workspace
- Physical risk factors
- Heat level/temperature
- Workspace layout
- Cleaning

- Air quality
- The number of workers – *crowding level*

There is no formal template to summarize the physical environment data collected, but it is certain that they can thus be lists, narratives, or visual descriptions of the users' physical environments you encountered or heard about, as exemplified above.

Social Environment Characteristics

A person's social environment includes their living and working conditions (e.g., roles and responsibilities), income level, educational background and the communities they are part of. In mobile user interface design, the social environment, either within the users' immediate group or in a larger group of co-workers and/or customers, may make a particular design difficult but acceptable to use. For instance, though texting is not easy to use, the elderly are keen to accept the mobile technology that their immediate family members are currently using, thanks to their social surroundings and environment (*a personal communication with a 67 year-old-lady living in Wellington, New Zealand, interviewed 23 July, 2006*). A user's milieu also includes his or her social positions and social roles as a whole that influence the individuals of a group.

A Guide for Practitioners:
The following items are to be considered in the social environment analysis:

- Income;
- Educational background;
- Level of collaboration;
- The communities that the users are part of;
- The number of workers – *their interpersonal relationship;*
- The number of users in the workspace;
- Attitudes and values to technology (special preferences);
- The level of understanding of the workplace or technology domain;
- Usage trends;
- Criticality of the system in social activities

Sometimes, the social environment of an individual is the culture that he or she was educated in and/or lives in, and the people and institutions with whom the person interacts. For instance, workers under pressure to perform functions quickly will be frustrated by interfaces that do not support them adequately. Members of the same social environment will often think in similar styles and patterns even when their conclusions differ[23]; that is, social environment refers to a person's relation with the society or community that he or she lives and works in. Since the onus in collecting and interpreting social environment characteristics is on the designer in providing appropriate mobile systems that can be used by the users under the specific social context, designing for the particular user group under the specific work context is generally considered an art. For one to see how to bridge an interface design with social environment characteristics, in developing a PDA-based "Supermarket" job dispatching system, our design team examined relevant social environmental characteristics of the workplace under the current work practices, in relation to the three intended user groups, as follows:

> The stocktaking staff are using the two main features of the computer system: stocktaking and product search. They need to record what products have to be restocked onto the corresponding shelf and other necessary details such as quantity, due time of stock or comments if any, mostly for the restocking staff.
>
> For stocktaking staff, prior working experience in a similar role is necessary. For the stocktaking staff to make an estimation of the number of products to be put back onto the shelves and by when they need to do so solely relies on their own experience. The number of items required on the shelf should not be greater than the total amount available for a particular product at the storage facility. On the other hand, one product may require a greater number than the other product because it sells faster. Thus, the quantity of the restocking items should be chosen carefully. This totally relies upon the stocktaking staff, so the skills below seem necessary:
>
> - Familiarity with the overall products the supermarket sells: Knowledge of a list of this kind of product will help staff improve efficiency and lead to increase in staff motivation in the long run.
> - Basic knowledge of information technology: The current system is a desktop computer system, so it is necessary for the staff to know how to use computers.
>
> The stocktaking staff are only able to check the products and to mark them in the system when it has to be restocked. Also, he or she can comment on what restocking tasks should be prioritized, reporting to the mangers.
>
> On the other hand, the restocking staff uses two main features: stock list display and product search. They have to use the computer system to check their work list and to change the status of tasks. For example, when a product is being restocked, its status should be changed to "In Progress" so the other restocking staff and stocktaking staff would know not to do the same job. It is thus necessary for the restocking staff to have the same level of information technology knowledge as the stocktaking staff to use the computer system effectively. Familiarity with the supermarket's internal layout will also greatly improve productivity for users in this category.
>
> The restocking staff can solely use the stock list display. This means that the restocking staff are able to see the list of products that have to be restocked. Moreover, they can report to the managers whether the product should be ordered or not and due to that it will be taken out of the stock list.
>
> The managers use this system to check all task progress in the supermarket. The managers are able to see all tasks with pending, in progress or completed status. They can read detailed information about a stock task, such as who has recorded it, how many items were required, which restocking staff member took up the task, when the task was completed and so forth. The managers also need to have more IT (information technology) knowledge to make using this system a productive experience. On the other hand, a highly experienced manager should know the processes of each user category (in particular, restocking staff) well so they can identify and deal with problems quickly.

---

[23] Common ground in electronically mediated communication: Clark's theory of language use (Monk, 2003) would be a good reference for understanding this issue.

Technical tips: Personality and interface preferences—Note that a popular technique to identify personality types is the Myers-Briggs Type Indicator (MBTI: Myers, 1993), which is based on Carl Jung's theories of personality types. Jung conjectured that there were four dichotomies:

- Extroversion vs. Introversion (where they prefer to focus their attention) – *the extroverts focus on external stimuli and like variety and action, whereas the introverts prefer familiar patterns, rely on their inner ideas, and work alone contentedly. The extraverted like face-to-face communication, such as video calls on mobile phones.*
- Sensing vs. Intuition (the way they prefer to take information) – *sensing types are attracted to established routines, are good at precise work and enjoy applying known skills, whereas intuitive types like solving new problems and discovering new relations but dislike taking time for precision. In particular, a sensing person needs a structured guide map for navigation or a highly structured task structure.*
- Perceptive vs. Judging (how they orient themselves to the external world) – *perceptive types like to learning about new situations, but may have trouble making decisions, whereas judging types like to make a careful plans and will seek to carry through the plan even if new facts change the goal. Hence, for judging personality individuals, the interface should be very simple to use.*
- Feeling vs. Thinking (the way they prefer to make decisions) – *feeling types are aware of other people's feelings, seek to please others and relate well to most people, whereas thinking types are unemotional, may treat people impersonally and like to put things in logical order. An aesthetical and personalized interface design is good for the feeling-type person.*

Cultural Environment Characteristics

Another perspective on mobile work context has something to do with cultural, ethnic, racial, or linguistic background (Katz and Aakhus, 2002). Differences between people within any given nation or culture are much greater than differences between groups. Education, social standings, religion[24], personality (see the technical tips above), belief structure, past experience, affection shown in the home, and/or a myriad of other factors will affect human behaviour. It seems obvious that users who were raised learning to read Japanese or Chinese will scan a screen differently from users who were raised learning to read English or the languages that read texts from left to right.

Users from cultures that have a more reflective style or respect for ancestral traditions may prefer interfaces different from those chosen by users from cultures that are more action oriented or novelty based. In Korea, particularly, "Yes" means, "I hear you" rather than "I agree." For interface design with this cultural background in mind, there should be differences in approach as to what is considered appropriate for "Confirmation." The user's cultural environment is not only about their ethnicity or nationality, but also an entire range of experiences related to their regional, professional, and socio-economic backgrounds. You may learn that you cannot use obtrusive sounds to signal computer errors in a culture where users may lose face if they think someone else knows they have done something wrong.

---

[24] LG™ Electronics announced, 16 September, 2008, the launch of the only TV with Qur'an built-in, to honor the Holy Month of Ramadan, cited in the AME Info 22 September, 2008.

Remember that your users may be a diverse group. The cultures you see depend on where you do the site visits. Even if you cannot visit all the cultures your product will reach, try to find out about those cultures from experts.

A Tool for Reasoning about Work Context I – Rich Picture

With regard to mobile work context analysis, simply interviewing or surveying users or workers at the workplace where interactions take place would be just fine. However, Monk (1997) proposed a representational technique to identify all of the stakeholders, their concerns, and some of the structure underlying the work context – a *rich picture*. A rich picture is a tool for recording and reasoning about these aspects of the work context—in particular, how they should affect the design. It is generally constructed by interviewing people. The ideal interview should take place at the workplace, as we discussed above, because the artifacts people use to do their work will be close at hand. It serves to organize and reason about all the information that users provide. Drawing the picture will point to places where you need to find out more or to apparent contradictions in the conclusions you have drawn.

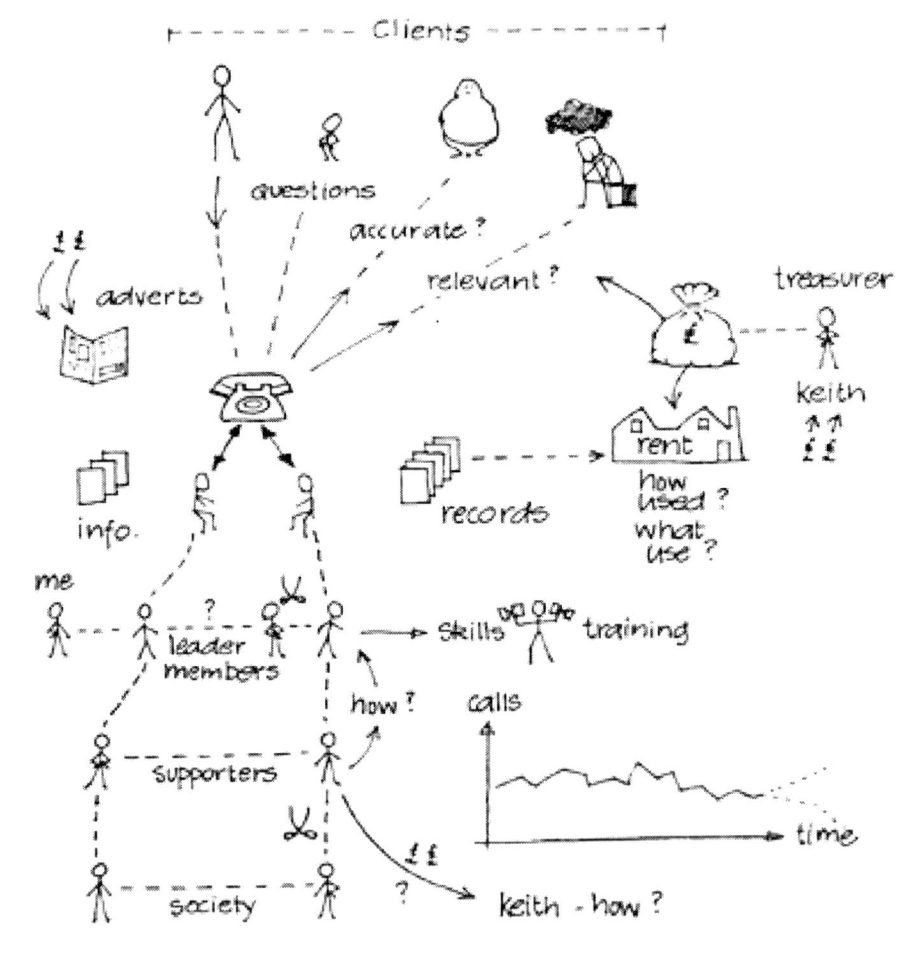

Figure 4.10. A rich picture, reprinted from Monk and Howard (1998).

The rich picture depicts the primary stakeholders, their interrelationships, and their concerns. It is intended to be a broad, high-grained view of the problem situation, serving as a starting point and a context for all of these activities (Monk and Howard, 1998). However, there is no single best way of producing a rich picture; the same analyst can use different styles under different circumstances.

The three most important components of a rich picture are *structure, process,* and *concerns*:

- *Structure* refers to aspects of the work context that are slow to change. These might be things such as the organizational hierarchy of a firm, geographic localities, physical equipment, and so on. Most important, it includes all of the people who will use or could conceivably be affected.
- *Process* refers to the transformations that occur in the process of the work. These transformations might be part of a flow of goods, documents, or data.
- *Concerns* are the most useful components, drawn in a thought bubble.

Drawing a rich picture requires that the analyst work closely with the stakeholders so that the picture captures the situation and related concerns from the stakeholders' point of view. Stakeholders can participate in the process by working with the analyst to identify structures, processes, and concerns significant to them.

A Tool for Reasoning about Work Context II – Activity Theory

Mobile devices, especially mobile phones, now occupy concurrent social spaces, in which users' mobile activities actually take place. This results in a lively public debate about what is acceptable and unacceptable in relation to mobile phone use in public spaces (of course, though this is highly associated with other elements of the mobile activity context). Now, many mobile phone companies are issuing guidelines on *mobile etiquette,* encouraging sensible and responsible mobile phone use behaviour in public spaces. Also, some train companies (e.g., Virgin™ train in the UK) now have *quiet carriages* where mobile phones have to be on silent mode and anyone wishing to make or receive a phone call is requested to go to the end of the carriage so as not to disturb the other passengers. Apart from this negative social position of mobile phones, mobile services can be designed to encourage mobile sociability in person as well as online *sociability*. In 2008, JuiceCaster™ launched a location-based geo-tagging service. With the geo-tagging service, JuiceCaster™ users can send pictures and videos that automatically include their location to various social networking platforms, including MySpace™, Facebook™, Bebo™ and Twitter™. Through its new location-based feature, JuiceCaster™ now allows individuals to see a video's location information, which can be used as a friend-finding service (i.e., mobile sociability). It also immediately notifies users when someone updates their status in a nearby area.

In designing such context-based mobile services, *activity theory* (AT: as a key reference, Nardi, 1997) has been appropriated as a righteous approach for identifying and relating the elements that should be taken into account in system development. In particular, in the sense that sociability is a key issue in the modern mobile usage trend as exemplified above, the application of AT has much been appraised to reason about other elements in the mobile work context. For instance, the author has recently modelled an activity system with AT about extramural students who worked on their course projects together, as depicted in Figure 4.11.

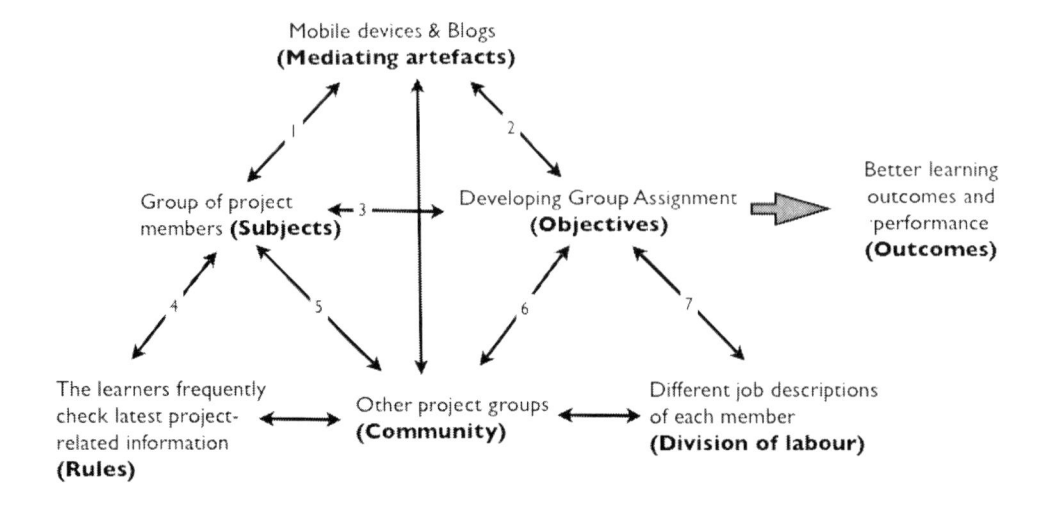

Figure 4.11. Activity is an intertwined system: if one entity changes, the whole system becomes unstable and must develop to obtain renewed stability. This is a part of modelling the relation among the six entities from extramural students' perspective.

This activity system draws a context for extramural students who had enrolled in the User-Centred Design (UCD) course at Massey University in conducting three group assignments. This course necessitated the group-based practical experiences to apply the UCD concepts and techniques from the online lecture for a more realistic systems design exercise, which is clearly represented as the learning outcome.

Principally, the group of extramural students were using two mediating artifacts (numbered "1" in the figure): *text messages* on their own mobile phone and *Web blog* systems. These mediating artifact-producing activities included the intrinsic communication activities among the group members. Using these two artifacts, the extramural students could complete their group assignments (arrow 2), which indicates the object activity, i.e., three assignments. In effect, to carry out the three group assignments, the students would have to be in contact with one another using text messages and further motivate their collaboration with their own project blog (arrow 3). These group assignments required a lot of interviews, literature reviews, and focus group studies (arrow 7), and the responsibilities and roles were divided among the members of the group. The group members were asked to regularly report their own work-in-progress among themselves via the blog or text messages (arrow 4), which were abiding rule-producing activities. The status of the other group projects was also available through the course blog (arrow 5), which was of great value to motivate (or even regulate) their own project performance throughout the online course (arrow 6). The activities captured in Figure 4.11 were being constantly developed (at least not static) to match new needs or requirements. For instance, if the lecturer wanted to be informed about the current status of the assignments, he or she could add one rule-producing activity in the corner of the triangle to re-create a whole new activity system; consequently, the changed part of the systematic structure naturally generated relevant requirements as needed.

As one can see in Figure 4.11, by modelling these situated requirements with AT, we were able to increase the richness of the learning activities and make it possible to produce a more useful learning experience at the university level. In particular, the situational and contextual requirements in the interwoven activities can be systematically modelled with the useful mechanism that activity theory provides. The three entities – *community*, *rules* and *division of labour* – which denote the situated social contexts, allow one to contemplate relevant situated requirements in an effective way. See Bertelsen and Bødker (2003) and Uden (2007) for further detail.

Further, AT is impressively committed to providing food for thought to the interface designer, to articulate how the work context given would accordingly change current work practices from one to the other. Consider the following example:

> A small group of friends sit around a dinner table, talking about the events of the day and their friends. A phone rings. Two people reach for their pockets, and it's Peter's phone. He discovers a text message from his girlfriend, Jennifer. And he quietly chuckles. He dashes off a response, during which time he is near people. He re-enters the conversation as soon as he hits send.

With the example above, one can see that Peter is managing several *micro-contexts* simultaneously. First, his dinner companions (community context) provide a social context, long-term or immediate. Second, their current topic of conversation (division of labour context) might encourage acceptance or deferral of a message from Jennifer. Third, the larger physical environment – restaurant – guides expectations and provides another micro-context. Fourth, each mobile application (mediating artefacts) – voice, text, or content – provides its own micro-context. Finally, the personalities on the other side of the mobile connection – Jennifer in this case, an impersonal relation to Peter's friends – provide another set of micro-contexts. In other words, the composition of companions and the group's history and personalities influence call acceptance (rule or community acceptance context). The following diagram depicts these micro-contexts around Peter's interactions at the given time.

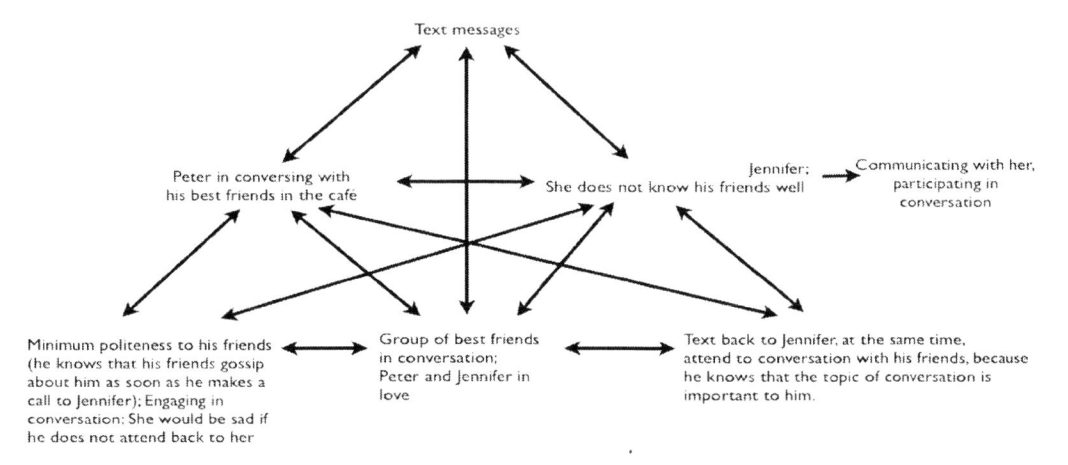

Macro-Contexts: Situation (conversing with his best friends), location (café), social object (Jennifer)
Micro-Contexts: He is dining with his friends (community context), He is seeing her (community context), Text messages (mediating application context), He knows three rules associated with his friends and Jennifer (rule contexts), He should attend to the conversation with his friends as well as text back to Jennifer (division of labour context)

A small group of friends sit around a dinner table, talking about the events of the day and their friends. A phone rings. Two people reach for their pockets, and it's Peter's phone. It is from his boss. He immediately answers the phone. The conversation at the table slows to a halt, with some people starting to look uncomfortable. Conversation slowly returns once Peter is off the phone.

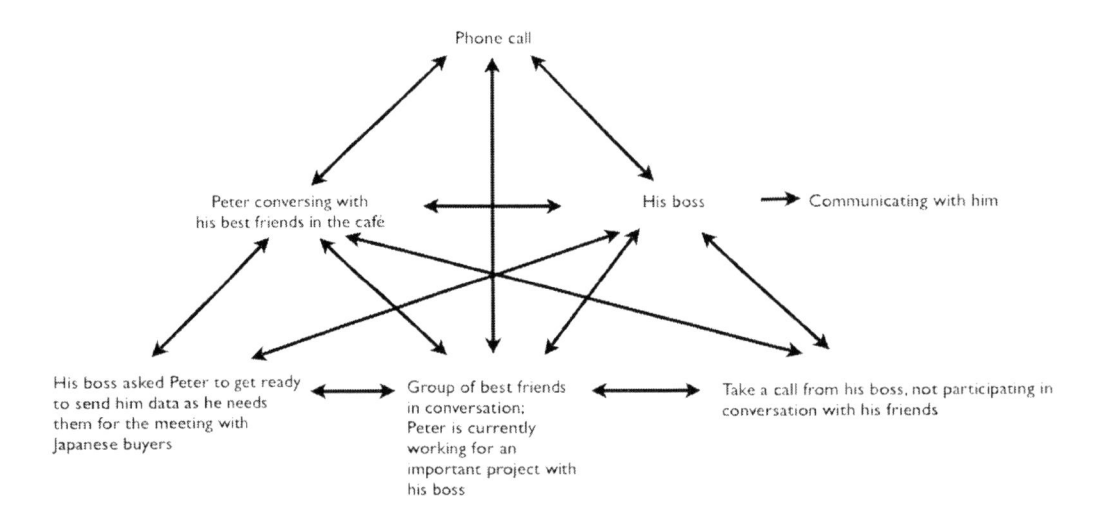

```
                              Phone call
                                  ↑
                            ↗    ↑ ↑    ↖
                          ↙      ↓ ↓      ↘
        Peter conversing with  ←→  His boss  →  Communicating with him
        his best friends in the cafe
              ↗ ↑                              ↑ ↘
            ↙   ↓                              ↓   ↘
His boss asked Peter to get ready  ←→  Group of best friends  ←→  Take a call from his boss, not participating in
to send him data as he needs           in conversation;              conversation with his friends
them for the meeting with              Peter is currently
Japanese buyers                        working for an
                                       important project with
                                       his boss
```

Macro-Contexts: Situation (conversing with his best friends), location (café), social object (boss)
Micro-Contexts: He is dining with his friends (community context), He and his boss are working together on an important project (community context), Phone call (mediating application context), He knows the rules associated with his boss (rule contexts), He should take a call from his boss, and not participate in conversing with his friends (division of labour context)

Contrary to the previous social context, in the scenario above, Peter interacts differently due to contrasting contextual distinction (i.e., different micro-contexts). Personal and cultural practices provide some of the context in this activity system. Peter could defer Jennifer's call until later in the first scenario, because he knows that his best friends will gossip about him while he talks with Jennifer. So he could have had his phone on silent, and answer her back via text messages. In contrast, once he has a phone call from his boss, he could not help but to answer it. That is, Peter is coping with several different micro-contexts against Jennifer's case.

As such, AT provides a lens through which to look at the surrounding micro-work contexts. It can provide useful broad provocations for the design team, leading to questions like *"Have we really understood the importance of this tool?"* or *"Have we really considered other elements in this whole socio-technical system?"* The primary focus of activity systems analysis is the production of some outcomes, which involves a *subject*, the *object* of the activity, and the *tools* that are used in the activity, as illustrated in the two scenarios above.

Subject: the subject of any activity is the individual or group of actors engaged in the activity. The subject (the user in many cases) is the central and driving character in defining activity. In other words, depending on the nature of the activity, the subject may change. Understanding the motives, conflicts and interpretations are the most important step of the process because it helps the designer understand and address the underlying dynamics that drive the rest of the system.

Object: the object of the activity is the physical or mental product that is transformed, or mostly the person who is interacting with the subject. It is the intention that motivates the activity. In the examples above, Peter's girlfriend and boss are the cases.

Community: this normally presents the social context that the activity systems run. The examples above, "group of best friends," are the common entities defining the social context of their response to Peter when the phone rings, and the relationship with the callers (Jennifer in the first drawing, and the boss in the second drawing) dictates Peter's responses. The community may oppose (in the first drawing) or support the activity or at least remain neutral (in the second drawing); it may facilitate or impede access to resources.

Mediating tools: the three primary components – *subject*, *object* and *community* – do not act on each other directly. Rather, their interactions are intervened, or mediated, by other factors. *Tools* mediate the relationship between the subject and the object, and can be anything used in the transformation process. The user of culture-specific tools shapes the way people act and think. Tools alter the activity and are, in turn, altered by the activity. If a specific tool is not available, subjects may adapt something else to use in its place, changing the way that they interpret or view the tools. As a result, how the task is approached and how it unfolds or develops will change. This may, in turn, change the new uses for the tool to perform that activity in the future. *Rules* mediate the relationship between the community and the subject. And *division of labours* (or *roles*) mediates the relationship between the community and the objects. These artefacts are both a result of and result from the interaction of the primary components. Rules are the mediators or negotiators between the subject and the community. These rules are expressed as culturally accepted norms for behaviour. For example, professional cultural norms are expressed in the ethical practices that are generally accepted within and among colleagues. Division of labour can be conceived as roles, which mediate the dynamic relationship between object and the community. Roles can be thought of as those activities that the community rewards.

Yet, the lack of practicality is always a concern in applying AT to systems design. To embrace pertinent situated contexts in HCI design processes, Bertelsen (2004) proposed *activity checklists* that seem to be usable in comparison to AT itself, as follows:

TECHNICAL TIPS: ACTIVITY CHECKLISTS (extended from Bertelsen [2004])

Means and ends

- *Who is using or will use the proposed technology?*
- *What are the goals and subgoals they are trying to achieve?*
- *What are the criteria for judging success/failure of goals?*
- *What troubleshooting strategies and techniques are used?*
- *What constraints are imposed by the goals on the choice and use of technologies?*
- *What potential conflicts exist between goals?*
- *What potential conflicts exist between the goals and goals associated with other technologies/activities?*
- *How are conflicts resolved?*

Environment

- *What other tools are available to users?*
- *How might other tools be integrated with the new technology?*

- *What level of access do users have to necessary tools?*
- *What degree of sharing is involved with tools?*
- *What is the spatial layout and temporal organizational of the environment?*
- *How is work divided up? Think about synchronous and asynchronous work between different locations.*
- *What are the rules, norms, and procedures regulating social interaction and coordination related to the use of the technology?*

Learning/cognition/articulation

- *What parts of the user's actions are to be learned and internalized?*
- *What knowledge about the technology remains "in the world"?*
- *How much time and effort is needed to master new operations?*
- *How much self-monitoring and reflection goes on with the users?*
- *How well are users supported in terms of being able to describe the problems they have with the tools?*
- *What strategies and procedures do exist to allow users to help each other when problems arise?*

Development

- *What effects might the proposed technology have on the way activities are carried out?*
- *What are users' attitudes toward the proposed technology and how might these change over time?*
- *What new goals might become attainable after the technology is implemented?*

The hardest part to understand in user's work contexts is the fact that the contexts are not static but more dynamic, which changes throughout a user's interactions with the surrounding world over time. While this purports the benefits of activity theory, we should also acknowledge that the generalization of the context is almost impossible, whether that can be illustrated by activity theory or not. Therefore, this context should be used either as excluders or as filters. Consequently, the design of interfaces for mobile devices needs to be fine-tuned by this contextual understanding to flesh out or carve an appropriate mobile interface design.

A PRACTICAL PROCESS OF WORK CONTEXT ANALYSIS

A work context analysis reveals who a user is and what he or she wants under what environmental circumstances. Then, more than one work context analysis would reveal what they are required to do (or what they would like to use the system given) both in the current and in the future (or newly developing) system. It thus ensures that you know who you are developing your products for, who to recruit for usability activities, and where to test them. The work context analysis is thus an analysis of what is currently performed by users and what features (such as their physical characteristics, cognitive requirements and their community to allow such activities and so forth) should be considered in systems design. It also results in documentation of the current users' viewpoints (if an updated system is being proposed).

To do work context analysis, all interface design begins with an understanding of the intended users, including population profiles that reflect age, gender, physical abilities, skill levels, education, cultural or ethnic background, training, motivation, goals, and personality. There are often several communities of users for a system, so the design effort is multiplied to cover the diverse user groups. The process of getting to know the users is never ending because there is so much to know and because the users keep changing. Every step in understanding the users and related contexts and in recognizing them as individuals whose outlook is different from the designer's own is likely to be a step closer to a successful design. In this section, we will focus on user profile analysis to capture its general process as part of the work context analysis, but the all/but same process will apply for environment profile analysis

For user profiles, first you need users. It is vital to get the right user groups; otherwise the data you collect will indicate only the subset of the intended user groups. The way to collect the right user groups is not so straightforward; however, the scope of the intended user groups can be obtained from the marketing department or customers support department, which have more user-related data from their working experiences. To systematically cover the diverse user experience levels, I normally categorize the intended user group into the three, as follows:

- *Novice users* – true novice users are assumed to know little of the task or interface concepts. This group of users may arrive with anxiety about using mobile devices, which inhibits learning. Overcoming these limitations is a serious challenge to the designer of the interface, and affects instructions, dialogue boxes, and online helps. Restricting vocabulary to a small number of familiar and consistent concept terms is essential to begin developing the user's knowledge. The number of actions should also be small, so that novice users can carry out simple tasks successfully and thus reduce anxiety, build confidence, and gain positive reinforcement. Informative feedback about the accomplishment of each task is helpful, and constructive, specific error messages should be provided when users make mistakes.
- *Knowledgeable users* – this group of persons have stable task concepts or knowledge, but they will have difficulty retaining the structure of menus or the location of features. The burden on their memories will be lightened by orderly structure in the menus, consistent terminology, and high interface transparency, which all emphasize *recognition* rather than *recall*. Consistent sequences of actions, meaningful messages, and guides to frequent patterns of usage will help these users rediscover how to perform their task properly.
- *Expert users* – expert users are thoroughly familiar with the task and interface concepts and seek to get their task done quickly. They demand rapid response time, brief and non-distracting feedback, and the capacity to carry out actions with a few keystrokes or selections.

Novice users can be taught a minimal subset of objects and actions with which to get started (the *training wheel* [25] concept by Carroll and Carrithers [1984]). They are most likely

---

[25] Training wheels are originally the kind of wheels you put on a child's bike when he/she is learning to ride a bike. A training wheel is a program/device/system that disables or hides advanced features so novices can learn the

to make correct choices when they have only a few options and are protected from making errors. After gaining confidence from hands-on experience, these users can progress to levels of task concepts and the accompanying interface concepts. On the contrary, both knowledgeable and expert users are to some extent familiar with the given or a similar system, so their expectation of the new product is quite different from what the novice user would have.

Creating a work context document is an iterative and reflective process. When you will likely have some idea of which user group would be your intended user category, you can start collecting their user profiles. However, note that as you conduct activities of work context documents and learn more about the end users, you should come back to the start and update it. This is the so-called *reflective process* in the work context analysis. You may not be right in guessing who would be your target user groups at the very first stage. Your first guess may be slightly off centre.

Your list of user groups and your character matrices may be enhanced by creating detailed user profiles of representative users. Here, do not simply define all stakeholders as "user"; instead, try to categorize them into one of two or three categories: *primary*, *secondary* and *tertiary* (if necessary). Primary users are those individuals who use and work regularly with the system (stocktaking and restocking staff in the example above); secondary users will use the product infrequently or intermittently. Very few cases find the tertiary user category, but certainly a large institution would have such a user category. In designing a mobile phone for kids, the primary user group would be the kids who are eager to use it; but the secondary user group (i.e., parents) would also need to be consulted for phone settings (such as parental guidance).

User profile documents can be lists or narratives or visual descriptions of the users you have visited. Checklists can be very convenient for defining relevant user characteristics. The objective in user profile checklists is to identify and document explicit characteristics that particular user groups have in common and which must be taken into account when making interface design decisions. Note that some of those listed below may be irrelevant in certain contexts and should be excluded accordingly. In a similar vein, if you need more, you should put them in the checklist. Figure 4.11 shows a sample user profile checklist.

To enhance your user profiles analysis further, you can include a narrative or visual description. The power of narrative descriptions and pictures of users is evident in that they tend to be more memorable to your team members than simple lists[26].

Once you determine the range of responses for each of the user characteristics, which contains the number of features you may be interested in, then you need to summarize the user profiles into groups based on their similarities (Courage and Baxter, 2005, p. 47). Some groups you may use are:

- *Age* (e.g., child, young adult, elderly)
- *User experience* (e.g., novice, expert, highly experienced with other systems)
- *Attitudes* (e.g., first adopter, technology-phobic, preference for other systems)

---

system faster in a protected environment where experimentation is safe and encouraged. Their goal is to save users from the frustration and confusion caused by the errors they make in the early stages of learning. That way, the user would have the ideal environment for building a coherent mental model of the system, resulting in better performance and learnability of the advanced functions after the initial "training wheel phase."

[26] Please note that the same accounts will be applied to environment profile analysis.

- *Primary tasks* (e.g., task-oriented, fun-seeking, hedonic, utilitarian)
- *Use trends* (e.g., loyal to the system, migrating from other systems)
- *Cognitive characteristics* (e.g., multi-task performer, multimedia-phobic)
- *Communities* (e.g., highly reliant on social activities, solitary)

User profile checklist

Use this form as a guide for collecting user profiles. The criteria listed below help you identify what information you need from users. However, please bear in mind that the items below are neither extensive nor comprehensive.

Date: ... / ... /

Any data obtained from this experiment will be made anonymous before being published. If at any point you wish to withdraw from the evaluation, you can do so and all your data will be destroyed.

	Demographic characteristics		Comments
1	Age		Categorise the age group rather than collecting the actual age of the user
2	Gender		
3	Location		
4	Socio economic status (related to Occupation)		

	Physical characteristics		Comments
1	Any disability?		Speech/hearing/visually/mobility impaired? Arthritis or shaking hands? Weak eyesight and so forth?
2	Handedness		Right/left/ambidextrous?

	Occupation		Comments
1	Current job title		
2	Years in this job		
3	Responsibilities		What roles are you currently doing in this organization?
4	Previous job titles/years/responsibilities in the job posts		Please detail them, including the roles that you had performed in the organizations.

	Language		Comments
1	Native spoken English?		If the system is for a different language (e.g., Chinese or Japanese), please modify this question
2	Fluent in written English?		The same with above

	User Experience		Comments
1	Educational background (degree, major, courses and so forth)		

Confidential
Hokyoung Ryu's Questionnaire
Date Published: 7/25/2005

Figure 4.11. (Continued on next page).

User profile checklist:

2	Mobile device use experience (years of experiences and skill levels)		Novices/knowledgeable/expert?
3	Attitudes and values (special preferences)		Are you an early-adapter or so?
4	The level of understanding of the product domain (including technologies)		Do you have any product you want to purchase in a near future?

	Specific User Experience		Comments
1	Experience with other mobile devices?		Use of other systems, if so what are they and how frequently you had used them? Have you used other similar systems? If so how many systems have you used before? One or some? Name them please
2	Usage trends (how long and frequent)		How long have you been using the product you are currently using? How frequently are you using the system?

	Tasks		Comments
1	What is the primary tasks?		
2	What is the secondary tasks?		
3	Why do you need the mobile device and for what?		
4	What tasks you like the most with your system?		
5	What tasks you like the least with your system?		

	Viewpoints		Comments
1	Do you think the system you are using can be better off the others?		Why?
2	Do you have any product that can be better off the product your are currently using?		Why?

	Cognitive styles		Comments
1	Are you an active learner?		
2	Do you prefer visual or audio contents?		
3	Do you like customizing the product?		
4	Can you recall your mum's telephone number now?		This question is very domain-specific for me to understand digital amnesia.
5	Are you often using your mobile phone while driving?		

User profile checklist

	Communities			
1	How many phone numbers do you keep in your mobile phone book?			
2	Do you normally ask the phone numbers when you meet new persons for your own business?			
3	Criticality of the system in social activites			Do you think the system is very important to manage your social activities? If so why?

General Observations[1]:

[1] This part should be filled by yourself in collecting this profile, if required.

Confidential Page 3 7/25/2008

Figure 4.11. An example of a user profile checklist.

The level of *user experience* is most important; experts hardly need coaching. On the contrary, novices are the opposite. For instance, if a mobile device you are designing is most likely to be used by hauler drivers (see the example below), one should note their cognitive requirements while driving.

In fact, there is no hard evidence of the best serving information that should go into a user profile checklist, but any information that helps to specify the capability of users to handle mobile devices, and to understand information presented in the device, is beneficial. This information may also be collected from reading market survey documents, observing users of similar systems or systems to be replaced, from interviews or other techniques that will be outlined later. In addition to generating profiles on current classes of users, we may need information about projected future users, which is another consideration that you should bear in mind in selecting the participants of user profile analysis.

User categories	Roles	Descriptions
Primary	Hauler in Greater Auckland area (User)	Any person using in-car navigation system or considering acquiring the device
Secondary	Hauler company dispatcher (Buyer)	People considering importing items into this country and effectively delivering them to stakeholders
Primary	—	People with some kind of physical disability (wearers of contact lenses, e.g.)

Figure 4.12. An example of a user profile summary.

As you have collected as many user profiles as possible, you need to summarize them as exemplified in Figure 4.12. The example is not definitive or even complete. It is intended to give some picture of the relevant user groups one typically needs to specify and show how one figures out who should be given user profile checklists.

Unlike the example above, which describes the two user categories by their roles, the following Figure 4.13 handles the same user profiles in different ways, by their viewpoint of the current system being used. This user profile analysis seems to determine what and how much new interface design is required.

User categories	Roles	Descriptions
Loyal users	Hauler company dispatcher (Buyer)	They like to keep using this system, because it took a long time to get used to it, though it was very difficult to use at first.
Hating the current system	Hauler in Greater Auckland area (User)	Whenever they have messages from the dispatcher, they first need to pull over their trucks because they cannot read such a small text while driving.

Figure 4.13. An example of a user profiles summary.

THE TECHNIQUES

This section discusses practical techniques with which you can collect data of users and their contexts. Please note that these techniques can also be employed in the other design stages, i.e., Chapter 5 (work analysis), and Chapter 10 (testing). All of them describe and analyze how one can collect the features of people, task, and work environment with various emphases. These may be grouped in many different ways, in my view, as follows:

- *Observation;*
- *Interview* – informal, semi-formal, structured, electronic interview, and focus group;
- *Ethnographical study* – diary and technological biography, field study, and cultural probe;
- *Formative description* – critical incident method (CIM)

Observation

Observation (frequently, in association with ethnographical study) is often used in mobile systems design, such as for gaining an understanding of how tasks are normally performed in actual settings. When we want to know what tasks, contexts, and environments are considered by the users, we need to observe some workers and record the tasks they expect to be completing.

The observation is all about seeing the action in its setting. To this end, the emphasis is on capturing as much of the detail as possible, and this involves taking notes, drawing quick sketches, photographing, and making audio and video recordings. However, there is a strong tendency for people to react – whether favourably or unfavourably – to being observed and recorded, in a way that they would not be if merely being measured in an apparently more impersonal way. In particular, where social interactions are a significant influence, the introduction of an observer will always have an effect, although this may wear off after a time. For this reason, one of the most important characteristics of an observational technique is the extent to which it intrudes, or appears to intrude, on the user's sense of *privacy*. It is important to first explain the technique to participants, and bear in mind that you need written permission to do so.

Observational techniques are most appropriate when the information of primary interest is of a visual or audible form. Thus, observation methods can be particularly useful for recording physical task sequences, or verbal interactions among several people. In cases where a high proportion of the task involves covert mental processing, and very little value is added by overt behavioural information, it is probably not worthwhile to perform observation. However, it may still be very useful to record verbal reports (see verbal protocols in Chapter 10) in addition to any objective performance measures that may be needed.

The following is part of the transcript of *artefact walkthrough* [27], which observes how a user employs a mobile device for her own goals. It appears to easily elicit some insight into a likely new key layout in something as simple as a verbal utterance.

---

[27] The purpose of artefact walkthrough is to review a user's responses conducted during the development of just about any kind of product.

* * *

[User A] [Action-Search] Key(Navigation) below the screen.

[User A] [Talking aloud] "well ... I am just thinking that this key may be better positioned in the middle of the keypad. ..."

[User A][Action-Hit] Key(Navigation)

* * *

By observation, we cannot legitimately say why the action was performed, as we do not have access to the person's planning and justification for an action on the basis of simply observing them. To cope with this problem, we may wish to record some commentary made by the person on their action. This could consist of a description of what they are doing and thinking, an account of their planning and intentions, and justification of what they have done, all of which can be considered *verbal protocol* (see Chapter 10 for further detail). A point worth noting, however, is that the use of verbal protocol may simply be a means of substituting one set of interferences concerning the cause of behaviour with another set. This could raise a question of whether or not we need to supplement our observations with information gathered via verbal protocol. A further problem arises from knowing what to observe and how often to observe it.

Types of Observation

There are two primary types of observations – *unobtrusive* and *obtrusive*. In unobtrusive observations, the observer does not interact with the person being observed. The observer tries to read from the situation so as not to change it or interfere in any way with the actions of the person being observed. In obtrusive forms of observation, the person doing the observation interacts with the person performing the tasks. In particular, an obtrusive form of observation should be sought when you:

- seek details about how complex or mental operations should be done;
- need to know why a performer did something;
- want to observe a few people for extended time; and
- intend to observe them again.

Obtrusive observation has several advantages and disadvantages against the counterpart.

Advantages

- It reveals information that cannot be acquired in any other way.
- It is ideal for pilot studies, as they can reveal potential behaviour patterns and influences that might not have been predicted.
- It allows the designer to become more familiar with the task, through experience.
- It can be used to identify and develop explanations of individual differences in task performance.

Disadvantages

- Situations that produce good or "context-rich" observational data are rarely the ones that produce precise and controlled data.

- Observation cannot provide information about underlying thought processes, and so they will be of little use for highly cognitive tasks.

A Guide for Practitioners to Perform Observation

- Before conducting any observation, it is essential to try to predict what information is expected to be extracted from the data.
- A pilot session is invaluable for assessing practical problems of data capture and the subsequent quality of the data.
- Explain the technique to participants, and then ask for written permission to do so.

Questionnaire

This is a very common tool used for gathering data in the area of HCI, and a fast and effective way to collect information about a service, application, and interface that they are currently using. Of course, it can be quantitatively analyzed, so the understanding from this technique may be more attractive to designers who are developing a mobile interface for a highly competitive market.

It consists of set of predetermined questions arranged on a form and typically answered in a fixed sequence. In the user interface context, questionnaires typically allow an investigator to either directly probe specific aspects of a task, or to examine the attitudes and feelings that may be expressed towards particular aspects of the task. In particular, it is useful for probing different perceptions and task knowledge of individuals with different backgrounds, either in terms of their previous use or experience of a system. For instance, differences between recommended procedures and actual action procedure could be investigated by comparing the responses of two different user groups.

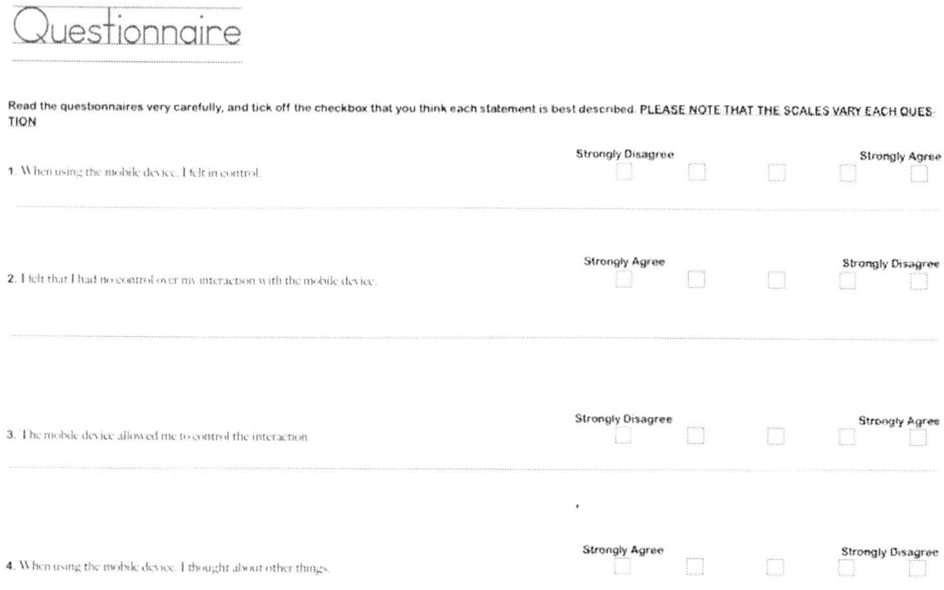
Questionnaire

Read the questionnaires very carefully, and tick off the checkbox that you think each statement is best described. PLEASE NOTE THAT THE SCALES VARY EACH QUESTION

1. When using the mobile device, I felt in control. — Strongly Disagree ☐ ☐ ☐ ☐ ☐ Strongly Agree

2. I felt that I had no control over my interaction with the mobile device. — Strongly Agree ☐ ☐ ☐ ☐ ☐ Strongly Disagree

3. The mobile device allowed me to control the interaction. — Strongly Disagree ☐ ☐ ☐ ☐ ☐ Strongly Agree

4. When using the mobile device, I thought about other things. — Strongly Agree ☐ ☐ ☐ ☐ ☐ Strongly Disagree

Figure 4.14. An example of questionnaire.

In constructing the *Likert-scale* questionnaire, it is important that you try to have an equal number of positively-worded and negatively-worded statements. This is important in order to control for the *acquiescence effect*,[28] a phenomenon whereby participants in a study may unwittingly try to respond positively to every question in order to help the investigation with their response. Or, alternatively, they may respond negatively to every question if they do not like you or your study, or simply to quickly finish off the questionnaire.

There are two practical ways to control this undesirable effect. Firstly, you can repeat similar questions in the questionnaire design; say, the first question is "The mobile phone was easy to use" and the fifth question is "I felt awkward while using the mobile phone." Consequently, consistent responses to the similar questions would be reliably considered as the nature of their attitude and feeling about the mobile phone. Secondly, the mixture of the order of "Strongly agree" and "Strongly disagree" (i.e., scales) would help to remove the *acquiescence effect*, as shown in Figure 4.14. This technical tip forces participants in the questionnaire study to intentionally think about what they are answering rather than blindly ticking off either "Strongly agree" or "Strongly disagree" to every statement on the same scale. In addition, it is of value to mix up the presentation of your positively and negatively worded statements in order to minimize the risk what is known as *constant error* partially caused by the *acquiescence effect*, as mentioned above.

A Guide for Practitioners: How to Design a Questionnaire for User Data Collection

- Questions should be kept short – *usually 20 words or less.*
- Avoid double-barrelled questions that address more than one issue at a time.
- Introducing multiple issues in a single question can be confusing.
- Vague questions can cause difficulty during an interview or questionnaires.
- Do not use double negatives.
- Do not use leading questions.
- Do not ask participants to predict the future.
- Use two similar questions to double-check whether they are consistently answering the questions.

Another type of questionnaire, unlike choosing the level of their agreement of the statement given in Likert scale, is the *semantic differential scale*. In a semantic differential scale an individual has to place a mark on scale between two bipolar adjectives based on how they feel towards the statement. For example, a question designed to evaluate the attitude of an individual might include the question "Do you think the mobile phone is very attractive?," and have a seven-point (or five-point) scale ranging from "very attractive" at one end to "very ugly" at the other end.

In relation to the types of questionnaire designs, there are a few issues you should be aware of. One problem is the mid-point on the scale. We should be very conservative to understand the neutral value when the participants choose this option. Quite often, the neutral value means that they are not really sure which direction they would rather go in. I generally take these neutral answers as missing values in the analysis without further interpretation. However, statistics-wise, the neutral point will be used as the reference face value for the

---

[28] A tendency to agree with the viewpoint of others, often with an authority. If the source is an authority, the acquiescent person will tend to toward agreement regardless of the nature of the content of the statement.

two-tailed t-test, as one quantitatively tries to analyze the data from the questionnaire study. See Ryu and Wong (2008) for further detail.

When you analyze the data from the questionnaire study, you may face *floor effect* or *ceiling effect*. The term *ceiling effect* is used to refer to the maximum score (e.g., five in a five-point Likert scale) on marking the questionnaire. The converse is called a *floor effect* (e.g., one in five-point Likert scale). In these two cases, it is quite difficult to generate any conclusive remark in the analysis, because statistically their responses do not include any random errors. However, whatever the statistical accounts are, the homogeneous pattern may imply the users' common attitudes or feelings towards the question stated, so it can be rather meaningfully interpreted.

The last thing you need to think about is the actual length of your questionnaire. I suggest that you keep your questionnaire down up to a maximum of about 20 statements. Please bear in mind that filling out the questionnaire is time-consuming for people, and this practice is at their will.

There are a number of standard questionnaires for conducting interface evaluation, though this is not the case in collecting work context data. One of the most popular is QUIS – Questionnaire for User Interaction Satisfaction (www.lap.umd.edu/QUIS). Many can also be found at Gary Perlman's site (acm.org/~perlman/question.html); however, none of them are specially designed for evaluating mobile systems.

Distribution of the questionnaire is normally accompanied by a self-addressed, stamped return envelope or in-person delivery. A more recent alternative to this expensive and laborious practice is to deliver the questionnaire over the Internet[29]. This has several advantages over using the previous telephone or mailed questionnaires. Clearly it is much faster and convenient. The respondents can respond to the questions very easily online and then just press the send button to return the questionnaire. However, the significant drawback of the Internet-based questionnaire would be the random clicking of their responses. Because the selection would be based on their free will, when they lose interest or willingness, they are often randomly selecting their answers to the questions. To avoid this problem, we may include two similar questions to double-check whether or not they consistently answer the questions. Alternatively, we can use different scales to collect a consistent answering pattern. For instance, the first question would use the scale from "strongly agree" to "strongly disagree," and the next question would use the opposite direction, as discussed above. Technically, Ryu (2003a) created a control procedure to avoid this likely random selection in his Internet-based experimentation: the participants needed to correct three times in a row to continue to the next experiment phase. Another possible drawback is that those without Internet connections could not be included in the sample population, thus potentially biasing the results as a consequence, by obtaining data from a restricted sample population.

As a concluding remark, if you are conducting a questionnaire study, you can begin by tallying the response to open-ended questions rather than closed-ended ones. It will help you consequently develop potential categories that can be used for your updated closed-ended questionnaire set, figuring out "what would be the range of responses you are getting to each question?" and "what would be the most frequent response?" When you analyze the data from the questionnaire, the statistical analysis would be the first choice, but regardless of the type

---

[29] As an interesting example, please visit ambysoft.com/surveys/agileMarch2007.html

of analyses, you can also select some illustrative answers to represent each category of response. Quite often, this will highlight the intrinsic nature of relevant user profiles.

A Guide for Practitioners to Design a Questionnaire

- Anonymity of respondents is a significant consideration.
- It is essential that designers fully appreciate what information they are seeking before they start formulating the questionnaire items.
- Try to obtain questionnaires that have been used in other similar studies to see how they have approached the problem and to see what they have used, such as QUIS (Questionnaires for User Interaction Satisfaction[30]).

Interview

The *interview* elicits a variety of information from a user or groups of users, which is especially useful for gathering unanticipated information about the interface and its surrounding environment, and is also useful when the mobile user interface designer is not completely sure of what the user wants to know about the tasks, functions, requirements, and contexts. Using interviews as a means of obtaining data concerning a user's knowledge and opinions of a system is an obvious tool to apply. Interviewing is used often in combination with observation (see above). One practical piece of advice concerns the amount of time that is reasonable to allocate for an interview. While this may vary by a person's availability, it is generally suggested that a 20-minute period will be a useful minimum, and 40 minutes' duration should be maximum.

When planning and conducting an interview, first the interviewer must select the interviewee who has the right access to the information the interviewer wants. That is, the interviewee has some background in the domain being explored, so the responses from the interviewee can be rightfully interpreted. This simple point is often ignored by many mobile user interface designers, so it is no surprise when a project fails in the end because of inappropriate user and environment profiles. Second, it is necessary to ensure that interviewees are motivated to give accurate answers. An interview that takes too long becomes tedious for people who want to get on with other things. Similarly, if the content of the interview is boring, too personal or unclear, the outcomes are consequently inaccurate. Also, for mobile work context analysis, in many instances, the interview needs to be administered at the workplaces in which mobile interactions actually take place. Finally, interviewer bias is a reality that must be carefully controlled throughout the whole process. *Interviewer effect* exists when interviewees respond in a manner that reflects what they think the interviewer wants to hear, rather than answers being an expression of the interviewee's own opinion. Therefore, interviewers must stop from showing any signs of disapproval, disagreement, or disappointment in the interviewee's response. For further detail on the interviewer effect, please see Chapter 10.

---

[30] Please bear in mind that QUIS can only be used for an evaluation purpose rather than collecting user profiles that is the main concern of this chapter.

Informal Interview

This is totally unstructured, with the aim of the interview being to try to collect as much data as possible. The interviewer can improvise the whole session with a small set of controlled procedures. That is, the interviewer does not begin with a firm agenda of questions or problems to solve. *Informal interviews* are usually more spontaneous and therefore more flexible than structured interviews (see below). They permit the exploration of a wider range of ideas and problems, such as brainstorming[31]. If you want to use an informal interview, you should employ this near the very beginning of the project to gain a broad perspective about the task.

Some always use a voice recorder during an interview to make sure to capture all that is being said. The data are then transcribed, and coding, analysis and interpretation follow from these records. The details of the technique to analyze this verbal record will be further discussed in Chapter 10. Others prefer to take notes, either because the voice recorder is perceived to be too intrusive and therefore to inhibit interviewees, or because they are certain they will get enough ideas down on paper to work from later.

Further, from the informal interview, a semi-formal interview is more directed to specific questions and question areas to be further explored. In order to stimulate further exploration during the semi-formal interview, probe whenever it is relevant. A probe is a word, phrase, sentence or utterance made by the interviewer in response to something the interviewee mentions.

Structured Interview (Formal Interview)

Unlike both the informal and semi-formal interview, all of the questions in a structured interview are fixed and set beforehand and there are a set number of responses that respondents can choose from when they answer the questions. However, structured interviews can also contain open questions. Though the time and effort required to do so is much higher than the other interview techniques above (i.e., informal and semi-formal interview), the results are said to be more reliably interpretable, because it is a directed information gathering process in which the interviewer controls the situations, sets the purpose of the interview, and controls the pace (Fisher, 1982). In this regard, the structured interview is considered the one that the interviewer uses to extract knowledge from the interviewee that is relevant to designing, evaluating, or exploring the possibility for creating a new mobile user interface.

> *A Guide for Practitioners: Structured Interview and Questionnaire*
>
> - The structured interview is very similar to a questionnaire study.
> - It is only conducted at a time when the designer knows precisely what issues are to be dealt with, when their boundaries have been clearly defined, and he or she is seeking quantifiable answers to well-defined questions while talking with users.

Electronic Interview

With the advent of Internet-based (or digital) communications, data collection has taken a new form, such as interviewing over the Internet, primarily video conferencing. Using Skype™, for instance, I have recently performed an interview session with a group of

---

[31] A method of shared problem solving in which all members of a group spontaneously contribute ideas, with little criticism of the ideas generated throughout the session.

Europeans to understand their different attitudes towards mobile learning applications. This seems to be useful, because

- Geographically distantly located users can be reached who could not attend a face-to-face interview.
- Interview scheduling and cost problems are minimized.

Contextual Inquiry

Contextual inquiry (Beyer and Holtzblatt, 1998) dictates that the designer should understand the users by interacting with them while and where activities are actually performed. There are several benefits arising from this *in situ* interviewing. The crucial difference with other interview techniques is that you must interview users in their real-life work context to understand their work and discover their user profiles and environments at the same time. Only then can you structure and present functionality or features of a mobile system in a way that taps into users' current work practices and optimally supports their current work practices.

Using this technique, we can guarantee a user model that represents the work from the users' point of view: how they currently think about, talk about, and do their work. The inquiries usually follow the following procedure:

1. The designer watches while participants begin some activity.
2. The designer then interrupts to question the reasons behind a particular practice.
3. Then the participant begins to provide a commentary, explaining while doing, like a master craftsman reflecting on his or her creative process.

A Guide for Practitioners to Conduct Interviews
(Extended from Courage and Baxter [2005])

- Before the interview
- *Identify what you want to find out.*
- *Choose the interviewees that are representative of the relevant population.*
- *Write the interview questions.*
- *Schedule the interview.*

- During the interview
- *Begin with an introduction – introduce yourself and explain the purpose of the interview and what will be done with the results. Make interviewees feel at ease by asking warm-up questions first, then the more delicate ones and, finally, a few routine questions at the end.*
- *Build a trusting relationship – show the interviewee that you are knowledgeable in the subject area. Ensure the interviewee that his or her expertise is important.*
- *Ask questions by topic – group your questions by topic. Keep the questions as specific as possible. Don't be afraid to go off topic if the information is valuable. If possible, limit the interview to 40 minutes to avoid any burnout.*
- *Listen to the interviewee – allow the interviewee to talk. Indicate your interest by nodding, commenting or body language. Probe the interviewee's comments with questions such as "Can you give me an example?," "Why would they do that?" or "What does the term X mean?"*

- *Take notes.*
- *Thank the interviewee for his/her time and effort. Also, remember the debriefing session, giving the interviewee a chance to ask questions.*

- After interview
- *Generate a post-interview summary – if you have taken notes during the interview, include explanatory material that may help in the later interpretation of the responses.*
- *Compile and analyze the results.*
- *Follow up – after reviewing your notes, you will always find some more questions to ask. Phone or email the interviewee with these questions, but minimize the amount. If you have sent a copy of the interview results to the interviewee, contact him or her to find out if he or she has any additions or revisions to the content.*

To conclude, two notes of interviewing are needed. Interviewing itself is a common technique for getting users to reflect on their experience in their own words, but it can also be used for testing interfaces (see Chapter 10 for further detail). The interview identifies mainly those areas one wants to explore further and other topics that one might want to confirm. It may take time to extract these topics from the data, but most often it is not too difficult to sift through the available data in search for the valuable parts. Finding statistical differences is, at many times, thus not significant.

Focus Group

Both interview and questionnaire have a major disadvantage. In the questionnaire study, the respondent is limited by the choices offered, and in both the interview and questionnaire study the findings can be unintentionally influenced by the interviewer or questions, by oversight or omission. In contrast, the *focus group* begins with limited assumptions and places considerable emphasis on getting in tune with the reality of the interviewees. Open-ended approaches allow the participants ample opportunity to comment, explain, and share experiences and attitudes, as opposed to the structured interview that is led by the interviewer.

Focus groups are typically composed of six to 10 people, but the size can vary from as few as four to as many as 12. The size is conditioned by two factors: it must be small enough for everyone to have a certain opportunity to share insights, and yet large enough to provide diversity of experiences. Small groups of four or five participants afford an opportunity to share ideas, but the restricted size also results in a small number of total ideas.

The most important point in recruiting participants of a focus group is that people have a variety of different experiences. The nature of the group is determined by the purpose of the study and should be a basis for recruitment. For instance, suppose an adult-community mobile learning program wants to know more about reaching people who are not currently participating in their services. In this case, the participants would be broadly defined as adults who live in the community who have not yet attended community-based mobile education sessions. Focus groups also have traditionally been composed of people who do not know each other, and thus can concentrate on the perceptions of the solutions, experiences, products and services. Hence, focus groups are not intended to develop consensus, to arrive at an agreeable plan, or to make decisions about which course of action to take.

Conducting a focus group study is a fairly unstructured group process. Krueger (1994) extensively explored many approaches to performing a focus group study, which is being increasingly applied by user interface designers to discover preferences for new or existing products. The focus group discussion is particularly effective in providing information about why people think or feel the way they do. In effect, the emphasis of a focus group is to shift attention from the interviewer to the respondent. Focus group study is not completed by a one-off meeting, and it is repeated several times with different groups of people. Typically, a focus group study will consist of a minimum of three focus groups, but could involve as many as several dozen groups.

In so doing, the designer must create a permissive environment for the focus group that nurtures different perceptions and points of view, without pressuring participants to vote, plan, or reach consensus. Therefore, the discussion can be comfortable and often enjoyable for participants as they share their ideas and perceptions. Group members influence each other by responding to ideas and comments in the discussion. Careful and systematic analysis of the discussions provides clues and insights as to how a product, service, or opportunity is perceived.

A Guide for Practitioners to Perform Focus Group Studies

- Do not allow criticism of ideas.
- Encourage quantity of ideas rather than being concerned initially with the quality of ideas.
- Verbalize ideas as soon as and as often as they occur.
- Provoke others by requiring them to state their assumption – this is your main responsibility (as a mediator) in the focus group study.

Ethnography

Ethnographic approaches to HCI and product development were pioneered by Alladi Venkatesh in the 1980s. Venkatesh pointed out that there might be wide gaps between what technologies can provide and what users actually want. Using longitudinal surveys and ethnographic field studies, Venkatesh's work (1996) examined various social contexts in which technologies were embedded. Venkatesh's approach led to rich insights into the context of technology use such as user resistance to new developments and the possibilities of new kinds of social interaction facilitated by technology, which seems to be quite relevant to new mobile systems design.

Ethnography was originally proposed in the design context as an intention of two changes: the development of new technologies and a growing realization that there was a need to understand the *user context* in which products and technologies are actually used. For the latter, ICTs (Information and Communication Technologies) become mediated among users, leading interest from traditional interface design issues to spread from understanding the single user at a desktop to understanding social interaction and work organization. At the same time, for the former, there was a need to move beyond the designer as subject to understanding the people who use a product or technology in their everyday lives. This arose partly because of the need to differentiate products, and partly because some design organizations recognized that they needed to know more about the users of products in order to create a better design. Interest in ethnography intensified in the Internet era in the late 1990s as technologies reached into the home and extended into areas of people's lives beyond the workplace. The last few years have seen a particular interest in mobile technologies as to how they have presented a new opportunity to work or enjoy activities together.

Jones (2008) rightly points out that ethnography's basic principle is the study of activities in their everyday settings, which is motivated by the following tenets:

- Only by understanding the context in which people live can we fully understand their activities and therefore their present and future needs and desires.
- People have only limited ability to describe what they do and how they do it without immediate access to the social and material aspects of their lives.
- Some aspects of people's experience can only be understood through observation.
- It frames a situation from an insider's view, which is often very different from an outsider's view.
- The outcome is a description of people's everyday realities; it does not in itself prescribe new practices, ways of working or new services and products.

The tools and techniques used in ethnography typically include: *observation, interviews*, and *self-reports*, such as *diary* studies and *visual stories*. Ethnography is not simply a set of data collection techniques; its value is gained by reflecting on a deep understanding of people: their cultural and symbolic frameworks, their activities and values. It is critical to sift through the data and to analyze the findings in order to identify the insights that demonstrate its genuine value.

Ethnographic methodologies have been used in conceptualizing future product requirements too. In Stolzoff et al. (2000), for instance, two interviews about smart home applications were conducted with each family at six-month intervals. Participant observation was also employed to examine interaction with the applications. Using the Venkatesh (1996) categories of physical/architectural space, social and cultural space, and technological space, they focused on current technology applications and attitudes toward smart home technology. They found that little had changed even in new buildings regarding physical and architectural space, implying that new smart home applications would not change the current work practices much as opposed to what the designers had imagined. Likewise, there have also been many ethnographic studies of mobile device uses (e.g., Esbjörnsson, Juhlin, and Weilenmann, 2007).

In ethnographic studies, the investigator attempts to be as unobtrusive as possible, finding a group of people already engaged in some interesting behavior. Most critically, the ethnographer adopts the position of uninformed outsider whose job it is to understand as much as possible about the "natives" from their own point of view. All else flows from this basic precept:

- The need to use a range of methods including intensive observation, in-depth interviewing, participation in cultural activities, and simply handing out, watching and learning as events unfold.
- The holistic perspective, in which everything – belief systems, rituals, institutions, artefacts, text, etc. – is grist for the analytical mill; and immersion in the field situation (Monk, Nardi, Gilber, Mantei, and McCarthy, 1993).

The main features of ethnographic research are as follows:

- You should have a clear idea of the particular social phenomena you are interested in observing.
- Focus on a small number of users.
- Analysis will be based on describing and explaining the human actions you have observed in a particular social context.

It is the author's personal belief that the climate becomes more suitable for the introduction and adaptation of ethnographic methods in collecting user or environmental profiles, which are designed to not only learn about others' environments, but also to do so from their perspective. An approach to studying users that makes use of information obtained by observing and talking with them in their natural social and work environments is important in designing artefacts for them.

Ethnographic methods are particularly useful for gathering information from users when little information about them is known to produce a prototype or an effective questionnaire. The following three techniques are often noted in the ethnographical study; otherwise unobtrusive observation is a normal technique.

Diary Study

As a special form of ethnographical study, a diary study asks participants to keep a personal record of their thoughts and feelings about interacting with a particular piece of mobile technology.

The key thing to remember here is that you have to give participants in your diary study specific instructions of what you want them to describe and write about. You should bear in mind that you cannot be with your participants.

A diary study is a self-reporting technique (see Figure 4.15) that helps fill in the gaps when you cannot observe them. In writing the entries, a participant can be given the opportunity to reflect on their experience, without the pressure or influences that they might feel when being observed.

To carry out the diary study, you should provide your participants with a way of recording their entries, and give them instructions on the sorts of observation you are interested in, the period over which the diary should be kept, and the frequency of entries you would like.

A Guide for Practitioners: Diary Study Questions to Participants, in Mobile Phone Uses. (Extended from Jones and Marsden [2006]).

- Please list the number of text messages that you made today and who you sent them to.
- Please list the number of mobile calls that you made today
- Please list the duration of each call and who you made the call to.
- Please list each mobile phone service that you used today and what you think of it.
- Please list the locations where you used your phone and what you used it for in these locations.
- If you used your phone in a public space, write down how you felt about using it in each of these places.

Time	Activity	Technology	Concerns
12am	Sleeping	...	
1	...	...	
2	...	...	
3	...	...	
4	...	...	
5	...	...	
6	Waking up		
7	Turning on TV	TV	Finding remote control
8	Reading newspaper Drinking coffee	Computer Coffee maker	
9	Driving to work	Car	Not much fuel
10	Sending email	Email client	
11	...	...	Too much typing
12pm	Lunch	Web search	Find the best place for today's lunch
1	...		
2	Call to client	Mobile phone	Search business cards
3	...	...	
4	...	...	
5	...	...	
6	Back to home	Car	Stop by gas station
7	Turning on TV	TV	
8	Dinner	Cooking recipe	What to cook tonight
9	...	...	
10	Go to bed	Alarm clock	Difficult to see whether it is armed or not
11	Sleeping	...	
12	...	...	

Figure 4.15. Timed diary activity log template.

Furthermore, the *technology diary*, as shown in Figure 4.15, can be an eclectic procedure in association with participants' critical and creative responses to the questions below. It allows one to see current uses, problems, and concerns, and by implication and elicitation of past developments and historical trends the factors that are of personal importance to the respondent, which will lead to desirable future development.

A Guide for Practitioners: Technology Diary Questions

- Who uses this most?
- Who uses this least?
- When do you use this?

- How often do you use it?
- When did you buy it?
- Why did you buy it?
- Have you had any trouble with it?
- Do you enjoy using it?
- How could it be improved?
- Do you wish it could do anything else?
- Are you planning to buy anything else soon?

Cultural Probe

As another form of diary study, the *cultural probe* (Gaver, Dunne, and Pacenti, 1999) provides a way of gathering information about people and their activities. The main difference between a diary study and a cultural probe is that an aspect of the cultural probe is adapted to elicit three wishes. These three wishes would be highly useful for the designer's awareness of users' current needs or wants.

To collect these three wishes, stamped, addressed letters were given to respondents along with three pieces of paper with the words "I wish I had . . ." written on each page. Letters rather than postcards were employed to reduce feelings of self-consciousness. In my latest collection of these wish lists from 68 respondents, wishes include voice-recognition technology to minimize text entry with mobile phones, a mirror for make-up, a way to find the mobile phone when it is misplaced, and the ability to use the mobile phone as a universal remote control.

Here two notes about ethnographical studies are needed. Ethnographical studies are conducted either in an effort to describe what is going on in a given environment or situation, or as field experiments. They aim to understand procedures or interactions between people as they normally occur when researchers are not present to record events. In these kinds of studies, the researchers behave unobtrusively, trying to "blend" in with the background so that his or her presence does not distort or hide the patterns of human communication he or she is trying to elucidate and understand.

Also, recruiting is most important, since the studies rely on a large investment of participants' time. Participants are expected, on average, to spend at least several hours during the course of the activity. Therefore, it is particularly important to ensure that you recruit the right participants, and that you monitor and support them as well as possible throughout the process.

As these two concerns cannot be guaranteed, in practice, it would be more appropriate for the researcher to become acquainted with the users' environments. Because it is very difficult for people to describe exactly what they do and how much time they spend on various activities, a pragmatic way to find out is for the designer to become part of the procedures using the so-called *participant observation* method.

If the people concerned know they are being studied, they are likely to put forward their best behaviour, work faster, and be more accurate, co-operative and conscientious than under normal circumstances. In this case, it would be appropriate for the researcher to join the team and perform the tasks in question herself/himself.

A Guide for Practitioners to Perform Ethnographical Studies

- Spend as much time as possible in the users' context.
- Be both in their world while at the same time retaining an objective, questioning perspective.
- Record as much of what is going on as possible.
- Take a field notebook.
- As well as watching, do some semi-structured interviewing in situ, if necessary.

Formative Approach

More favourably, the *formative approach* reveals requirements that must be met in system design so that the system could behave in a new and desired way. It is used to identify a job or task profile by isolating and prioritizing the behaviours that are essential to the job or task. Hence, it is very effective to identify users' needs that the users are little aware of. In particular, the specification of projected design attributes would be effectively publicized. One of the favourite techniques in so doing is the *critical incident method* (Flanagan, 1954), by which designers are well aware of the criticality of errors or ideas of a new task design.

Critical Incident Method

The *critical incident method* was developed to elicit particular stories that exemplify extreme situations. It employs a semi-structured interview format with specific, focused probes to elicit goals, options, cues, contextual elements, and situation assessment specific to particular situations. The kinds of information that are normally collected about each incident reported should include the following:

- Circumstances leading up to the incident;
- Description of what the person did;
- Why the incident was helpful/detrimental to the goal of the person;
- When the incident occurred;
- Description of the person's job;
- Assessment of the person's experience level in the job

The analyst next distills this information into statements of critical competence. These statements represent the critical incidents or competencies required by any task or users.

To gather the incidents, the critical incident method relies on the survey method of interview. You are asking respondents for their impressions or attitudes about the critical elements, presenting the following questions:

- Describe an incident you remember that was an example of an effective (or ineffective) intervention.
- What were the general circumstances leading up to this incident?
- Tell me exactly what the _____ did that was so effective (or ineffective) at the time.
- How did this incident contribute to the overall goal or effort?

Then, the incidents, which include contextual and attitudinal information, need to be summarized (or documented) into useful statements that can be analyzed further. Please note that the incidents that any individual recalls depend on the memory of the respondents. Therefore, the incidents are likely to vary. So these incidents often need to be further analyzed for their criticality. At this stage, they probably should be grouped together or organized in order to minimize the effects of response set bias. This involves three sub-procedures. First, you must select an appropriate frame of reference for describing the events. Next, select a set of headings for classifying the events. This is usually done inductively. Card sorting[32] would be a good technique in this case. Finally, the levels of generality or specificity appropriate to the analysis must be determined. The actual arrangement would depend on the purpose of the analysis.

Figure 4.16, reprinted from Ryu and Parsons (2008), illustrates the process of collecting the university students' learning experiences by using the *critical incident method*. We first collected as many critical and practical incidents as possible from 20 university students, asking what types of contexts, tasks, information and design features would suit a university student's expectations.

After transcribing the interview data, we decided upon a coding scheme containing two categories for handling the interview data: *mobility* issues and *learning* issues. In the first category, the interviewees presented some requirements relating to location issues around the campus. In the second category, the interviewees presented several ineffective learning experiences that they had at the university. The frequency column in Figure 4.16 gives the number of interviewees who contributed to each category. Note that one asterisk (*) indicates excerpts from the first-year students, and two asterisks (**) indicate excerpts from those who are senior students (i.e., second and third year students).

Interestingly, the frequencies in Figure 4.16, as derived from the protocols, appeared to show certain differences between the two intended user groups (new students vs. senior students). That is, the senior students had more concerns about learning issues, whereas the new students reported more mobility issues. These interview data served to specifically identify the design requirements of the mobile learning application in terms of three perspectives. Firstly, the interview data revealed that most of the new students (seven out of eight students) had little idea of where the classrooms and laboratory facilities were located in a widely distributed campus (in fact, there are three separate precincts at Massey University in Auckland). Because of this, new students are often unsure of where their next meeting or lecture is to take place. In contrast, many of the senior students seem to be already aware of this type of information, but require more in-depth information about their personal studies (i.e., "learning issues'). Their concerns during "campus life" revolve around the organization of their studies, such as being aware of assessments, while also being up to date with messages and resources from lecturers and/or other students involved in their program of study. The new students' contexts and requirements were thus relevant to aspects of interface design, such as map support, while the requirements of the senior students related to the personalization of their learning environment.

---

[32] Please see Chapter 6 for the technique.

Category	Frequency New students	Frequency Senior students	Sample excerpts
Mobility (or location) issues	7	1	*I am a new student, so sometimes…, I have no idea of where QA2 is, I don't even know what QA2 means. That means I missed the first lecture in my first-ever school experience. * The lecturers never told us where to go for the tutorials and lab sessions. I was looking for the labs immediately after the lecture, but it wasn't so easy to find the place, which is in the other precinct, in five minutes. ** This … course changed the lecture room in Week 5. But I didn't get any notice of this room change (interviewer asked why). Actually, I missed the Week 4 lecture, so I also missed the Week 5 lecture.
Learning (or contents) issues	2	10	* I want the university to help me easily find assignments, examinations, library information, room changes, and so on, cos' the lecturers never kept the course outlines, so I have some difficulties finding the correct information. ** When I arrive on campus in the morning, I like to set up today's itinerary, where I have to go now, and then when or where I can have lunch, and so on. Also, if I have some time off between lectures, I want to stay at the library to read some books and do some assignments. However, sometimes the library is full of students, so I cannot find a place to study. ** The lecturer told us that we had quizzes next week, but I didn't know what I had to prepare for this, so I had been to his office, but he was not in the office. And then I found his telephone number from my mobile phone, and left a message in his voice mailbox. But he never returned my call. ** I requested an inter-library loan book. … I didn't collect this book for some time, because I didn't know I could. I had no calls or emails from the library, and they charged me a $2 fine.

* The excerpts from new students; ** The excerpts from senior students

Figure 4.16. Interview data summaries. Frequency indicates the number of interviewees from each user profile which gave the incidents for each category.

Another context found was the students' learning context, which emphasizes the needs and intended outcomes of specific learning activities. In particular, the interview found that spatial awareness is very important for the junior students, but temporal issues were more relevant to the senior students—for example, relating the time of day to their study schedules.

One of the interview responses showed that when a senior student arrives on the campus, the contextual information that they are searching for may be related to their course schedule for the day, room changes or important messages (e.g., assessment information, library records, and so forth). In this way, these different contextual profiles obtained from *the critical incident method* had defined different design requirements of a mobile learning system proposed.

Note that the critical incident method is of great use in identifying cognitive elements that are central to performing actions, but events are subject to error or rely heavily on biases/preferences and the accuracy of the memory of the respondent when the events are collected.

DOCUMENTATION

When all of the data, i.e., questionnaire, interview data and so forth, are returned or recorded, it is time to analyze data as planned and produce a data summary in a tailored format. The first thing to do is to write a short summary providing a synopsis of the key characteristics of each user category (*user profiles documents*) and their contexts (*environment profiles documents*), and draw specific implications for user interface design. It also summarizes the general needs of each different user category and variations within the category in relation to their working contexts. For example, if elderly people were found to be primarily interested in contacting their immediate family members only, the general needs of the elderly could be summarized as ease-of-use in calling a small number of people, and then examples of how to achieve these goals could be offered.

Lists or descriptions would be just fine in documentation. Alternatively, you can summarize the user and environment profiles in a table or spreadsheet, as depicted below. The following table shows the part of the document from the interview data in designing a mobile phone interface for the elderly. Hence, all of the participants were people aged 65 years and older, and a structured interview was employed.

Further, it can be condensed in related user categories; for example, the user category "the elderly with children" would have more interest in using text messages, and their most important concerns would be how easily they could send text messages without wearing glasses or other further assistance. If possible, you may include further statistical data to support your justification.

Question 1	Answer	Tally	Any special reasons [Note]
Enjoying using text messages?	Yes	////	Their children sent it first, so simply returned it.
	Sometimes	//	Not enough money...
	No	/	Too-small screen and keys are too tiny.
Question 2			
How many calls per day?	+5	//	Friends and relatives
	1–4	///	Relatives
	Never	//	No children to call
Question 3			
How many phone numbers?	+30	///	Friends and local police station
	10–30	///	Relatives
	<10	/	Children
Question 4	...	...	...
Question 5	...	...	...

As you are preparing these user or environment profile documents, you will naturally come up with several usability goals for a particular user category, e.g., *ease-of-use* for the elderly in this case. This is of great value in finding out what the proposed system needs to do with this particular user category and the contexts for these usability issues and so forth. For each usability goal, you can specify what it is, what its justification is and a metric that shows how the system can be measured against the usability goal, as exemplified in the table below.

User category	System must fit/do	System must avoid	Solutions
Elderly people	*Easy to use*	*Difficult to use*	*No conventional keypad; no screen; bigger buttons*
- Elderly with children - Elderly without children	Easy to find their phone numbers; easy connection to them	Small text	One-off customization
	Emergency contacts		Help line connection

Chapter 5

MOBILE WORK ANALYSIS

Every effort in mobile user interface design seems to make a mobile system easy to use. However, there is no hard evidence of what "ease of use" means, and what defines one system as easier to use than another. This is because ease of use is *personal* and *qualitative*, which is not at all easy to measure. Accordingly, designers tend to make a common mistake, being confused about *usefulness*[33] versus *usability*[34]. In fact, mobile designers have long been tempted to add more features or functions to their system design. It was not all that many years ago that mobile phones, for instance, had one basic function, and that was making calls. Over the past decade, though, more and more features and functions have been integrated into what was once a simple, one-task machine, to create what it is now commonly referred to as a convergent information device that would include every single device (e.g., a five mega-pixel still camera, camcorder, 60GB hard drive, satellite radio, audio/video player, PDA with Bluetooth™ and Wi-Fi and GPS with built-in antenna, and so forth). However, adding new features is not always a strength, particularly when existing users are confused and new users are discouraged from even starting. Arguably, a good mobile system is designed in such a way that the power of the system is tailored to support the maximum number of intended users to carry out the tasks that they truly want or need to execute. A poorly-designed mobile system wastes that power by providing unnecessary functions presented in unfamiliar and confusing ways. In a user-centred design, task (or work) analysis is thus an important and early component of the design process. It guides your mobile system design to remain close to what users actually want or need to do with the given mobile system.

The main outcomes of the previous design stage (Chapter 4, mobile work context analysis) are the answers to both "who are our users?" and "in what and where do they live or work?" Building upon them, the second stage in our design process is trying to answer the question "what do they want to do or need to do?" *Classifying*, *decomposing* and *rearranging* task components are equally important in mobile work analysis. The first – *classifying* – would be useful for structuring the tasks in a way that enables users to quickly use the system, and the other two – *decomposing* and *rearranging* – would be helpful in structuring the tasks in a way that allows users to accomplish the tasks as quickly as possible. Taken together with the work context analysis done in Stage 1, here, mobile work analysis proposes new work

---

[33] For something to be useful means that the user can actually achieve the task that her or she wants to execute with the given system (i.e., in relation to the quantity of features that the system provides).
[34] For something to be usable it must allow the user to achieve the task they he or she wants to execute easily and enjoyably (i.e., in relation to the quality of features that the system provides).

practices, as shown in Figure 5.1, by which the designer can determine what and how tasks should be realized in the product. This accordingly plugs into Stage 3, i.e., task-function mapping (Chapter 6), at a later time.

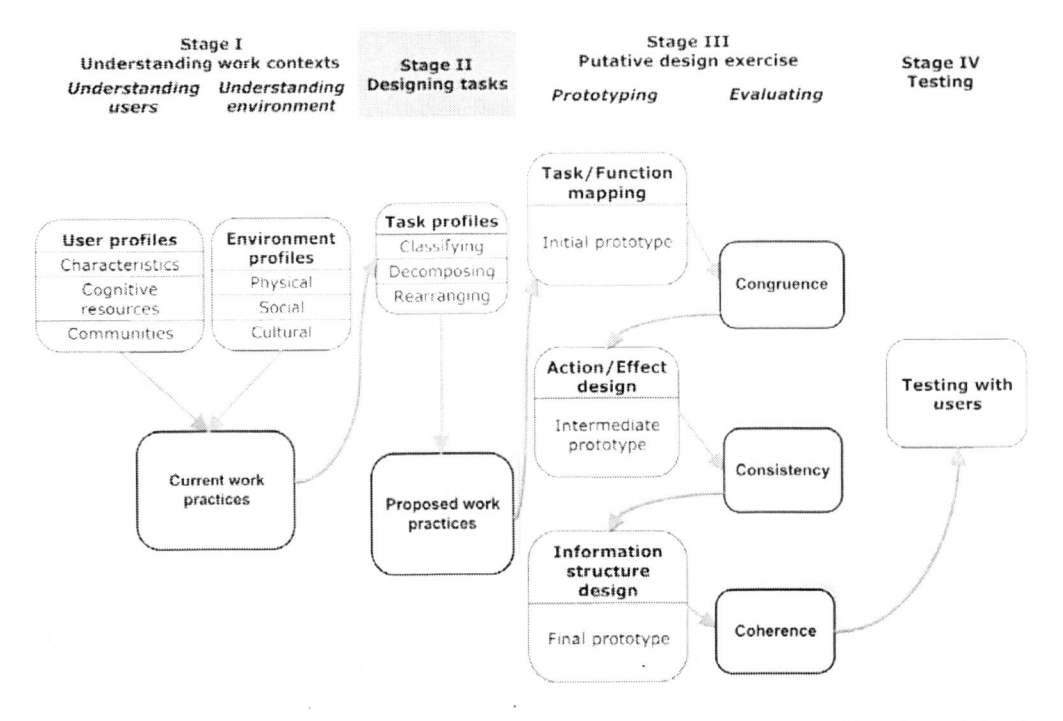

Figure 5.1. Mobile user interface design process: the outcomes from Stage 1 are fed into Stage 2 (work analysis), resulting in proposed work practices.

For the supermarket example discussed in Chapter 4, the primary tasks (current work practices) of the stocktaking staff are to:

- [*Intermediate cognitive tasks*] Record which products have to be restocked into the corresponding shelf and other necessary details such as quantity, due time of stock items or comments, if necessary.
- [*Highly cognitive tasks*] Make an estimation of the number of products to be put back onto the shelves and by when this should be done.

Compared to the two cognitive tasks being performed by the stocktaking staff, the restocking staff are performing a *physical task* (restocking the products) and a *low-level cognitive task* (when a product is being restocked its status should be updated to "in progress" or "complete" as it is done). In contrast, the mangers should be able to see all stocktaking and restocking staff members' jobs, categorized as pending, in progress or completed status, as their primary tasks (*supervisory cognitive tasks*). Building upon these current work practices of all the stakeholders, we were able to create new work practices with a PDA-based supermarket management system. This design exercise will be further discussed later.

Work analysis (Janassen, Tessmer, and Hannum, 1999; Kirwan and Ainsworth, 1992) involves the study of what the user is required to do to achieve a system goal. Hence, the primary purpose of work analysis is to compare the demands of a mobile system on the user with the capabilities of the user within their surrounding contexts (please note that these have been retained from Stage 1 – *work context analysis*), and, if necessary, to alter these demands, thereby reducing errors and achieving successful performance with the mobile system given. This process usually involves data collection of the task demands and representation of these data in such a way that a meaningful comparison can be made between the demands and the user's capabilities.

Work analysis for mobile user interface design is more than a process of analyzing and articulating the type of work that the users to know how to perform. If you are unable to articulate the ways that you want users to think and act, how can you believe that you can design mobile user interfaces that will help them? Mobile user interface designers thus perform work analysis in order to determine:

- The goals and objectives of the mobile device;
- The operational components of work—that is, describing what tasks users do, how they perform a task or apply a skill, in particular, relating to *task-function mapping* (see Chapter 6 for further detail) and *information design* (see Chapter 8 for further detail);
- Which tasks are more important – which have priority for a commitment;
- The sequences in which tasks are performed and should be learned and taught, in particular, relating to *action-effect design* (see Chapter 7 for further detail) and *information design* (see Chapter 8 for further detail).

In actual fact, *work analysis* is the most important part of our putative design process. It guides the following design stages by articulating the goal or mission for the subsequent design processes. We have seen too many mobile interface design projects fail to produce effective task procedures because the designers do not understand the importance of work analysis. For instance, my first mobile phone was somewhat tricky for me to send a text message, because I had to enter the recipient's phone number first and then proceed down to the next screen (note that these two task components were not displayed in the same screen) to compose text messages. Of course, though this personal experience is not hard evidence of a better task procedure of "sending text messages," at least it can pose the question, if not answer, of whether or not the designer had a second opportunity to more practically see his or her own design decisions.

Indeed, for HCI researchers, this is easier said than done. One of the predicaments of UCD is how to create a bridge between Stage 1 (work context analysis) and Stage 2 (work analysis). Specifically, many argue that the human elements must be incorporated into work analysis, and equally implicit and extensive are the human elements. This logically leads us to examine a systematic method, otherwise it is very unlikely that the human elements will be optimized, or that the potential for error will be minimized. Usage of explicit work analysis (i.e., *cognitive task analysis*) is therefore the instrument that allows effective integration of the human elements into system design and operations. It is also possible to establish a two-way flow of information in work analysis, with knowledge about human requirements and limitations feeding into the work analysis, and design preferences and constraints feeding into the work analysis. This can avoid the situation in which mobile phone design dictates user requirements or learning, which may lead to sub-optimal systems and the necessity of retro-fit design solutions later on. If work analysis is undertaken at the very initial conceptual phase of a system, there is the opportunity to address all human factor issues in the most cost-effective manner. It must be stressed, however, that work analysis is rarely a one-off process; instead, it usually requires one or more iterations as more detailed information about the system becomes established.

It is therefore most appropriate that mobile user interface designers undertake this work analysis to look at particular areas of concern, for specific benefits; yet, as mentioned above, it seems that work analysis is a kind of art that is most dependent upon the skills and expertise of the analyst. Therefore, it is widely said that work analysis is ambiguous. The ambiguity of work analysis also results from the myriad of contextual constraints imposed by the work context analysis performed in the previous design stage. Indeed, we claim that the purpose of work analysis is to obtain a user-centred model of work as it is currently performed, ultimately to propose new work practices. That is, as you understand how users currently think about, talk about, and do their work in their actual work environment, you can design the new system, finding optimal design solutions, proposing the best work practices to more effectively support users and users' existing task knowledge, and maximizing the current work practices by accommodating human cognitive constraints and capabilities within the context of their actual tasks.

To be fair, work analysis is uncertain, as not every aspect of human thought and behaviour can be identified or articulated. It sounds like an impossibility. However, the work context analysis performed in the previous chapter can generate some food for thought regarding work analysis by identifying categories of users whose tasks must be studied. Also, the current user knowledge and experience collected in the previous stage can be further exploited in this stage to facilitate proper user-centric work model design. And the outcomes from this work analysis are fed into the next design process—*task-function mapping*—where current work practices are re-engineered only as much as necessary to fit the nature of the tasks that users will be performing.

Although there is no universal consensus regarding the procedure in which work analysis functions are performed, we recommend the following examples as a general sequence that can be applied in most cases.

STEP 1. COLLECTING CURRENT WORK PRACTICES – GENERATING THE POOL OF TASKS

The order process starts when customers enter the restaurant, and a waiter/waitress leads them to an empty table (not a reserved one), and gives each one of them a menu. Just after that, the waiter/waitress asks them if they would like to have drinks (e.g., free tea or water) and brings over any necessary tableware before any meals have been ordered. Then the waiter/waitress will prepare the drinks and tableware for the customers, while the customers look through the menu to choose their dishes.

Next, the customers wave their hands or knock the table to attract the waiter's/waitress' attention, so they can come over either to take an order or answer questions.

There are three different models of ordering meals: first, the customers already know what they want to order. Second, they look for special dishes or special prices and order accordingly. Third, they don't know what exactly they want to eat; therefore, they just look through the menu and decide on the one they would prefer to have.

The customers can order dishes from any of the following categories: appetizers/starters, main dishes, drinks and desserts. Also, they can order meals based on different ingredients, including meat, poultry, seafood, and vegetarian. Customers can customize their dishes (such as the level of spiciness) and/or other services by telling the waiter/waitress to record their requirements on the order. When the customers order dishes, the waiter/waitress writes down the dish name, quantity, and the customizations (if any) on the paper pad with a pen. The order is then read back to the customers to make sure the order is exactly what the customers requested. Finally, the waiter/waitress leaves the table and takes one order slip to the kitchen and another copy to reception.

Waiters or waitresses work in the restaurant by leading customers to tables, taking orders, providing help, delivering food and drinks, cleaning tables, answering questions about dishes and drinks, etc. Waiters/waitresses record orders on carbon-backed docket books, sending one copy to reception and the other to the kitchen.

The first step is to collect current work practices, which are naturally plugged into what we have understood from the previous work context analysis. As illustrated above, the current work practices are mostly from the work context analysis done in Stage 1; however, quite often, determination of what tasks must be considered in the future product development is vague. Therefore, in this step we need to further fine-tune the relevant pool of activities or tasks to be included in the proposed system, based on users and environment characteristics (i.e., work contexts). In many cases, tasks to be included in the proposed system tend to be mandated or more tailored in advance by the other business sources, such as a customers service department.

STEP 2. CLASSIFYING WORKS (OR ACTIVITIES) FOR FURTHER ANALYSIS

It is important for the design team to investigate and analyze current system problems, because the technology in consideration may enhance the current workflows of the restaurant or cause the system performance to worsen. There are some problems in the current work practices performed in this Chinese restaurant.

1. Miscommunication: During the busy hours, waiters or waitresses frequently forget to record a customer's requirements, consequently taking wrong orders. More importantly, tourists, who have a language barrier, appear to have a poor experience with the staff. Also, some of the waiters and waitresses cannot speak English fluently, which can have negative impacts on the communication between them and the customer, resulting in occurrences of errors and omissions during the order taking process.
2. Slow order taking: In the current ordering process, each new order requires the waiters/waitresses to walk back and forth from the table to the kitchen to reception, meaning that a lot of time and effort is being expended with little return. Therefore, some customers may find it difficult to attract waiters'/waitresses' attention, and they may be left waiting for a long time to place an order or speak to a staff member. In addition, customers sometimes change their orders, and waiters sometimes forget these changes.
3. Unclear handwriting: In peak periods, waiters and waitresses write the order quickly by hand on paper, so the chefs and kitchen staff may prepare an incorrect meal, resulting in the order having to be re-cooked and the customer left feeling unhappy about the service

of the restaurant. The cashiers might also make errors while calculating the bill because of this unclear handwriting.

Having identified all of the tasks involved in the current work practices, it is obvious that there are too many tasks to analyze or to develop, so the pool of tasks or activities from the previous step needs to be classified in order to select the tasks to be analyzed in the subsequent analysis. However, in many design cases, the selection of tasks is committed prematurely, with inadequate experience, and is not performed systematically. In this step, the *performance criticality* and *knowledge states* required of users are classified as to the kind of tasks being selected. *Feasibility* and *priority* are thus the primary concerns here. Given limited resources, the analyst must evaluate the tasks identified in the inventory to determine which have priority in terms of *criticality* (most onerous or erroneous tasks), *frequency* (most frequent tasks), or *user preferences* (new tasks). We suggest that if the task (or the activity) is onerous (in relation to *cognitive workload*), performed frequently (in relation to *physical performance*), and/or newly proposed (in relation to *new information processing*), that should be associated with the subsequent analysis. As demonstrated in the excerpts above, we identified the three critical work activities that need to be appraised by the subsequent analysis: miscommunication, slow order taking, and unclear handwriting.

Quite often, mobile user interface designers have been forced to analyze all of the tasks in this step, by which they appear to be streamlined to skip this time-consuming process. However, in part most tasks are not totally new, and partly because many of the tasks have already been analyzed in the early product development they may be eliminated from the list of tasks to be re-analyzed here. Further, while designers are analyzing tasks in this phase, they become well aware of the issues or concerns to be taken into consideration, so they will attain the best practices for designing other tasks consequently. Therefore, you do not need to analyze all of the tasks, but to select them in terms of *feasibility* and *priority*.

STEP 3. DECOMPOSING THE WORK (OR ACTIVITIES) SELECTED, AND COGNITIVELY DESCRIBING THEM

Having decided what tasks are to be further analyzed or reviewed, the next step is to *decompose* them into subtasks or operations. However, it is reputable that this *decomposition* is necessary, in that processes such as *cognitive work analysis* rightfully focus on both the performance of the user and the internal knowledge states of the user, without the *divide-and-conquer* process which does not make sense, as the user would not necessarily perform as such. Nonetheless, my personal experiences tell that this decomposition process benefits *information design* (such as menus and structure of functions) in mobile user interface design, effectively dealing with a greater number of functions in the mobile system proposal. Of course, cognitive work analysis offers some advantages over decomposition-based analysis methods, but should be complementary rather than being put before decomposition.

In so doing, firstly, it begins by identifying relevant sub-components of the tasks selected. These hierarchies are developed by identifying what must be completed before each component should be acquired. The resulting arrangement of this process can be drawn in a diagram or chart[35]. This approach assumes that for performing an overall task, a set of subtasks exists such that when these necessary prerequisites are mastered, success is acquired in the overall task. Therefore, we can discover these prerequisite sub-task components through this structured decomposition process.

This decomposition process must be incorporated with a *cognitive elaboration* of the task components[36]. The emphasis here is *thoroughness* – ensuring the important task components with knowledge required to perform a task and its association with the objective of the task. This, in fact, is a primary rationale for conducting work analysis. Here, you must select one cognitive task analysis method or more that is appropriate to be able to *cognitively* describe the task components identified from the structured decomposition process. As you describe a task in both decomposing and cognitive ways, you are able to identify the operations (physical and mental activity) required to complete the task, the sequence of prerequisite tasks, or the constituent parts of a concept or principle. Thorough description of a task is important because you want to avoid omitting an important part of the task procedure in the next step – rearranging task components.

STEP 4. REARRANGING TASK COMPONENTS

Having broken down a task into its component parts and their cognitive requirements, next you need to rearrange them in a way that best conveys the task or that best facilitates using the task. *Rearranging* them is often naturally derived from the previous phase, i.e., as soon as you have decomposed and described the cognitive requirements of the task components. However, the task sequence is more than a simple enumeration of the components in which the task is performed. It should indicate the sequence upon which the user can naturally build. Cognitive descriptions are thus crucial here. Further, the sequence for performing the task implies an appropriate *information design*, which reduces cognitive workloads required to perform the task. In the previous decomposition step, the designer becomes aware of what components and their cognitive skills required are explicit, so this rearrangement step can help the designer develop an appropriate information design for the mobile user interface, such as feedback or other mode signals and so forth. Thanks to the nature of design, the outcomes of this rearranging process may create tension or contradict the current work practices collected in Stage 1, resulting in new work practices.

In so doing, *scenario-based design* (Alexander and Maiden, 2004; Carroll, 2002, 2003; Carroll, Kellogg, and Rosson, 1991) has been appropriated in many successful design exercises, representing every major use case.

The following scenarios were developed while we were designing the PDA-based supermarket management system, exemplified in Chapter 4.

---

[35] The details of how to do this are described in the section "hierarchical work analysis"
[36] Interaction Units scenarios are discussed in this book

A SCENARIO OF USING THE PDA-BASED MANAGEMENT SYSTEM: STOCKTAKING STAFF

Any shelf in the supermarket is accessed by its unique identifier, for instance, a location mark 'H3V4' means that the shelf is on the third aisle from the main entrance, and on the forth aisle westward. The work of stocktaking has been divided into sub-tasks according to the categories of the product such as Food, Clothes, Cleaning and so forth.

Sam is a new employee who has just had a 3-day training session before he was given the task of stocktaking fruits. He came to the Manager's office and got a PDA which has the "Shelf Stock Checking System (SSCS)" installed. He switched on the PDA by pressing the "Power" button on the PDA as he walked towards the shelves marked with "Fruits." Sam found the shelf with half empty. As Joe, an experienced staff member, has told him, this brand apple always sells well. He typed in his User ID and password from the virtual keyboard on the screen, and then clicked "Ok" to log onto the system. However, a warning box came up, saying that "The system could not log you on. Make sure your User Name and Password are correct and try again." He thought he might've typed in the wrong password, so he clicked "Ok" to close the warning, and re-typed the password.

A main menu with "Stocktaking", "Product Search" and "Log off" options was displayed when Sam finally logged in. He clicked the "Stocktaking" button, which led to a new screen. He found that there were only 5 bags of Sun Apple left on the shelf, and at least 55 more bags are needed to fill up the shelf. So he put in the serial number "7788257" by clicking the numerical buttons on the screen. When he finished entering the serial number and then clicked the "Next" button, a page with information of Sun Apple appeared. This page showed that 500 bags are currently available in the storage. When he decided to make an order he found that he was confused about how to do it. There are four buttons: "Make an order", "Return to the main Menu", "Back" and "Next", all of which seem available, except that "Next" greyed out. Fortunately, he remembered what his manager told him: "Go for 'Help' before coming to see me! ". So he clicked the "Help" icon and a help page is displayed, showing the information and operations available in previous page.

...

He pressed "Ok" to close the help page and clicked the "Make an Order" button as instructed. After that, the "Next" button was enabled, he clicked it and went on. He typed "55" into the text box and pressed "Confirm" to finish the order. It took a few seconds, and then the page jumped to the stand-by status. "Oh, so easy!" Sam cried. After he checked all the products, he went to the main menu to "Log Off".

A SCENARIO OF USING THE PDA-BASED MANAGEMENT SYSTEM: RESTOCKING STAFF

Danny's duty as a supermarket staff member is to restock the shelves. He was given a PDA with the "SSCS" installed to check what products have been requested for restocking. He logged into the system with his ID and password, and then recognised the options "Stock list display", "Product search" and "Log off". He clicked the "Stock list display" button. A list of products was displayed in the order of priority. The first product in the list was 55 bags of Sun Apples. Danny clicked this product and another screen came up to show the detailed information about this job. He recognised where these apples were in the storage and where they should be put back on the shelves. There were 4 buttons at the bottom of the system: "In Progress", "Completed", "Return to the main menu" and "Back". He decided to take on this task, so he clicked "In Progress". There's only the "Completed" button visible on the screen now. He went into the storage and carted out 55 bags of Sun Apples and laid these apples out on the shelf indicated on the PDA.

When the task was completed he then clicked the "Completed" button on the PDA. The screen took a few seconds to refresh, and then it went back to the list of products with KiwiSave Toilet Paper being the first product. He clicked the "Back" button to go to the main menu then clicked "Log off" button to log him out of the system.

A SCENARIO OF USING THE PDA-BASED SYSTEM: MANAGER

Paul is a manager in the local supermarket. His job is to check whether the supermarket's staff members do their job consistently and correctly. The vision of the whole supermarket is to guarantee that their customers will never find an empty shelf for a product that's not on clear-out sale. When Paul arrives at work every morning he started his day with a round of tour of the supermarket. In this first round tour he looks for shelves that haven't been restocked because all shelves should be restocked before the first customer enters the store. He walks through the store to check each shelf to see whether any one of them is empty. When he walks past the third aisle he realises that the shelf where the dog food should be is empty.

He switches on his PDA immediately, logs into the system with his ID and password and can see "Stocktaking", "Stock list display", "Management options", "Product search" and "Log off" options. He clicks the button "Management options" in order to enter the management function of the system. He enters "Logging view" by clicking the button "Log view", and then the log page is displayed. The "Return to the main menu" and "Next" buttons are at the bottom of the page. He is able to find the corresponding product quickly, since the log list is arranged in the alphabetical order of product names.

Paul clicks this task to see the status of the dog food restocking is "Completed" and Peter is the last person who marked it. There are 2 buttons at the end of the task: "Relist" and "Back". Paul clicks "Relist" and the page refreshes and goes back to the "Logging view". He then clicks the "Return to the main menu" button and goes to the main menu. He clicks the "Stock list display" button and the list of products needs restocking is displayed. The dog food that's been relisted is displayed.

Paul sees Dave walking by and asks him to restock the dog food. He also noted this incident down so he could bring this up in the next staff meeting.

Proposing new work practices with *scenario-based design* seems perfectly fine and correct; however, I feel that to some extent it has not fully lived up to the promise inherent in this book, which is a matter of art from the practitioner's perspective, so the focus of this chapter goes to a more practically engineered approach for the mobile user interface designer, in the hope that the designer would like to take the scenario-based design approach is being directed to the references above.

STEP 5. PROPOSING NEW WORK PRACTICES

It is important to note that proposing new work practices is the most innovative and creative step in mobile user interface design, so there are very few exemplars that are indicative of as part of an engineering process. My characterization is thus only a tentative one.

Both the decomposition and cognitive work analysis discussed above have formative implications for design. The very benefit of the combination of decomposition and cognitive work analysis is to be able to develop models of intrinsic work constraints for a particular task. The nature of the focus on intrinsic work constraints rather than tasks and devices is to overcome the problems with the *task-artefact lifecycle* (Carroll et al., 1991) – strong device-dependence and incompleteness. This combined approach will thus allow the properties of the device and the resulting work practices to emerge as outputs of the work analysis, rather than being assumed as givens.

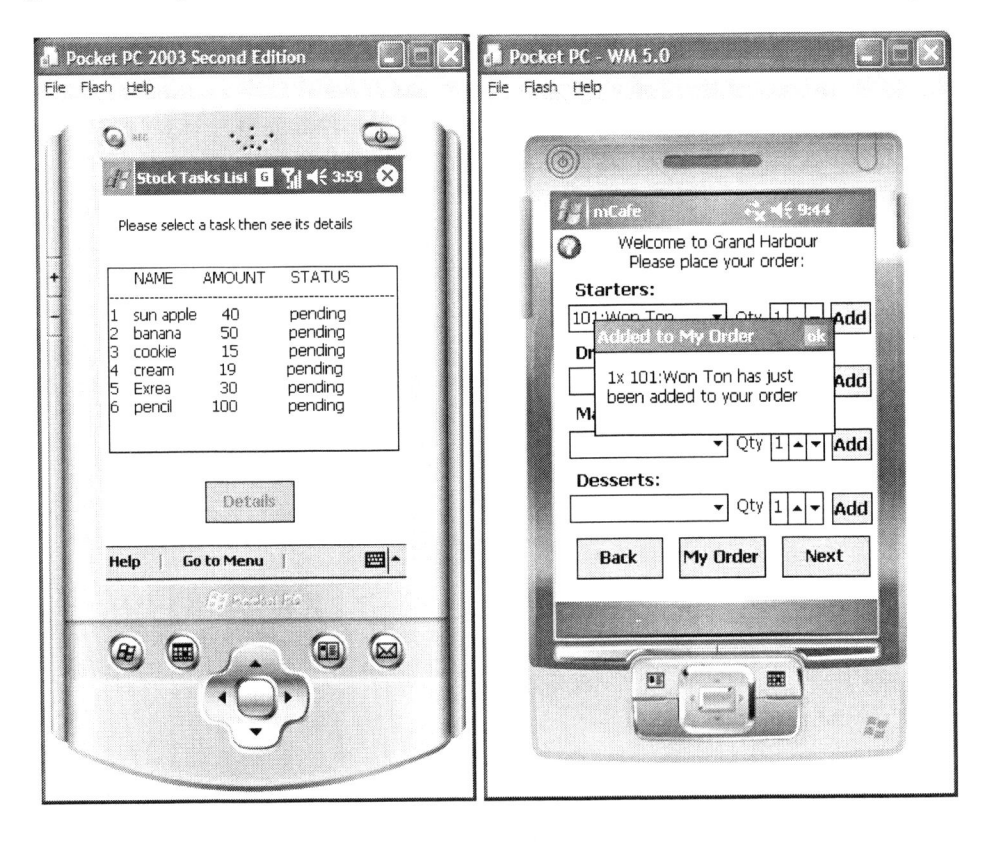

In the ideal case, this set of new work practices should correspond to the set of actions (or functions with the system newly proposed) defined by the intrinsic work constraints, thereby eliminating many inappropriate design alternatives and unexplored work possibilities. In the project that was exemplified above, we proposed new work practices for the restocking staff with a PDA-based system. Also, the diagram below illustrates their new work practices, and the screenshots above, show a first prototype for the re-stocking staff based on their intrinsic work constraints.

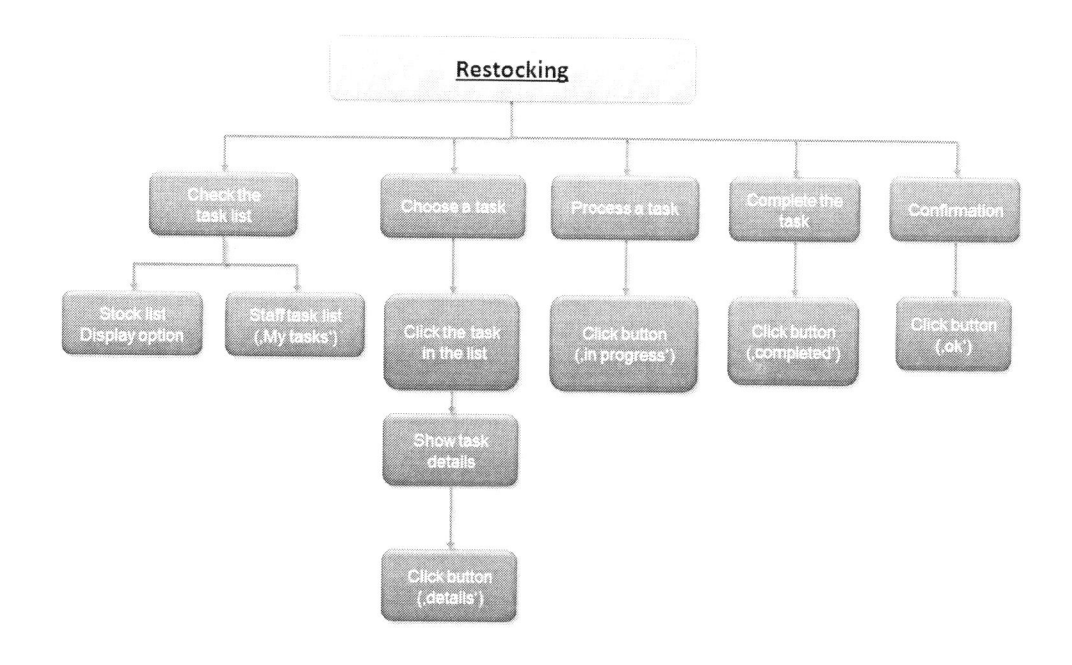

NEW WINE IN NEW BOTTLES

Perhaps the most ill-defined step in the work analysis stated above might be the last one, i.e., proposing new work practices, though it is the most important outcome to be plugged into the subsequent prototype design. Many practitioners have criticized its lack of systematic perspective, so many of them are merely undermining the power of work analysis. Again, a good work analysis is a step-by-step description of the demands an operation makes on the user with user capabilities (Drury, 1983), described in terms of operational procedures and knowledge necessary to complete a task (Benyon, Gree, and Bental, 1998). In effect, the primary purpose of work analysis is to gain an understanding of what people do in existing systems and what people can do with the proposed systems. Please note that I have not limited this two-flow process with the traditional term of task analysis here, in that the conventional purpose of task analysis normally refers to the former, i.e., what people do in existing systems. In our design process, the former has been mostly collected in Stage 1, so this work analysis stage further focuses on proposing new work practices with the system being developed.

Work analysis is a methodology which is supported by a large number of methods to help the designer arrange task components and to make explicit the requirements to be fulfilled by users and the proposed systems, and to optimize the capabilities of both parties. Therefore, work analysis is applied to make judgments about the best way to allocate human and machine resources for ensuring productivity, performance and satisfaction in human-machine and or human-human interaction. Hence, the outcome of work analysis can directly guide the subsequent design stage, *task-function mapping*, a first prototype design.

Underlying the notion of work analysis is that user's knowledge or cognitive states should be analyzed and modelled here to propose better work practices. In order to complete a task efficiently and systematically, people are assumed to use their mental processes, and these mental processes are assumed to be represented by information about goals, mental processes required, system feedback, and actions (Ryu and Monk, in press). A so-called *interaction unit* (IU) represents crucial system information and mental processes people have regarding performing tasks that they are currently doing (*recognition* or *affordance*) or have previously learned (*recall*). The IU model assumes that people use a new system based on previous knowledge (recall) or simply assuming the affordance of action-effect relationship, and apply them to task performance. It is a relatively systematic way for the designer to simulate users' responses in advance, and processes in a familiar or a novel situation. The IU scenarios provide a method for the analysis and modelling of tasks in terms of goals, natural action procedures with relevant mental processes (*recall-recognition-affordance*). However, it may be time-consuming to do the IU analysis, so the importance of criticality and representativeness of the tasks selected should be stressed again. That is why we take into account that the *classifying – decomposing – rearranging* three-legged process is necessary, prior to the IU analysis. In producing an IU scenario, the designer can identify what triggers the action, and what information (e.g., feedback or mode signals) is required to make the user take the action, so that one can systematically map the outcomes of work analysis onto the very first prototype design.

SELECTING WORKS (OR ACTIVITIES) FOR ANALYSIS

As shown in Figure 5.1, as Stage 1 is completed, the designer gets to know many of current work practices in different work contexts for different user categories. The understanding is forwarded to the second stage (designing tasks), and lays out new work practices. Here it is important to notice that we cannot analyze all of the tasks that the targeted users are to perform with a mobile system given. To ensure a systematic decision process of what tasks should be included in work analysis, four criteria – *difficulty*, *frequency*, *criticality*, and *novelty* would be of great benefit.

- *Difficulty* – how difficult is it to learn to perform the work or activity?
- *Frequency* – how frequently is it performed?
- *Criticality* – how important is the performance of the work to the goal or mission of the product?
- *Novelty* – whether the work given is new or not?

Next, these four categories take weighted measures on a 100-point scale to assign appropriate weight to each task, as shown in Figure 5.2.

	DIFFICULTY (10)	FREQUENCY (30)	CRITICALITY (30)	NOVELTY (30)	TOTAL	NOTES
Task 1 (Sending PXT)	10	10	15	25	60	
Task 2 (Sending TXT)	0	30	30	0	60	
Task 3 (Personalizing theme)	10	0	0	30	40	Newly added feature
Task 4 (Accessing the Internet)	10	0	30	10	50	

Figure 5.2. Task selection worksheet.

Assessing difficulty
Consider:
- Amount of time average users would need to learn the task
- The number of action steps to complete the task
- The proportion of the goals of the task completed
- How long it takes to complete a task

Rate the difficulty of each task 0–10 (0= easy, 10 = extremely difficult to learn)

Assessing frequency
Consider:
- The number of contexts where task is performed
- How frequently does a user have to access this task to complete his or her work task?

Rate the frequency of each task 0–30 (0= never performed, 30 = frequently performed)

Assessing novelty
Consider:
- Quite new?
- Can functions be understood by the user correctly?
- Can users understand what is required as input and what is provided as output?
- Can the user easily understand information from the current system?

Rate the novelty of each task 0–30 (0= conventional, 30 = new)

Assessing criticality
Consider:
- Safety or mission?
- What is the incidence of safety or health problems among users?
- What is the incidence of hazard to people affected by use of the system?
- Can the user easily correct errors in tasks?
- Is this task critical to the operational environment?

Rate the criticality of each task 0–30 (0= unnecessary, 30 = absolutely critical)

Figure 5.3. Technical tips of determining the tasks to be included in the work analysis.

In Figure 5.2, it can be seen that the two tasks, "Sending PXT" and "Sending TXT" have the same weight, which seem to be analyzed first, before Task 3, "Personalizing theme," which is a newly added feature. It can also be seen that "Accessing the Internet" is worth considering in terms of its criticality and difficulty. Indeed, one of my surveys with 68 university students (aged 19 to 25) revealed that "Accessing the Internet" was perceived as the most onerous task with their own mobile handset. Further details on the four criteria are depicted in Figure 5.3. Please make sure that the information published here is neither exclusive nor exhaustive; any information critical to the designer's decision should be included at his or her disposal.

Classifying all of the tasks based on the four criteria, the usefulness is obvious. The thorough assessment of systematically selecting current work practices would require a large-scale study; however, using this simple technique to select the tasks to be analyzed, one can be sure to have an opportunity to explore the most essential tasks or work activities first.

THE TECHNIQUES

Various techniques for work analysis are currently available in the HCI discipline, all of which describe tasks at different levels of detail and with various emphases (Kirwan and Ainsworth, 1992). Hence, user interface designers are often asked to perform several rounds of work analysis, which often gives us different looks through a different lens in speculating how works are performed in that light. Different goals and contexts require different approaches to consolidate new work practices; one size does not fit all.

First it should be decided what kind of analysis to perform and to learn how to select the appropriate method. There are, indeed, many methods for performing each, and then you must decide which of the many methods will produce the most appropriate outcomes for the given context. Each method for performing work analysis yields a different outcome that will result in a different kind of work model. It is thus important to keep in mind that the primary goal of all forms of work analysis is *proposing better work practices* (new task profiles). However, it is not quite proposed that designers must become skilled in every work analysis method. Rather, I believe that it is more important that user interface designers tailor their own work analysis approach and exploit the work analysis technique once a decision about the tasks to be analyzed has been made. In this book, I discuss two general classes of work analysis techniques that have emerged from my many years of design experiences: *hierarchical work analysis* and *cognitive work analysis*. For the former, a task decomposition method (*hierarchical task analysis*) is discussed, and *interaction unit scenarios* for the latter.

Not only do these two methods involve different procedures for fulfilling the purposes of work analysis, they also make different assumptions about how people do their tasks and so provide different recommendations for the proposed work practices. In a nutshell, interface designers are offered two possible criteria to choose one or the other:

- For information design, use a *cognitive work analysis*.
- For task procedure design, use a *hierarchical work analysis*, analyzing job steps and articulating sequences of tasks.

Hierarchical Work Analysis

HTA (Duncan, 1972) is a technique that defines a user's goal in a top-down fashion as a series of actions or sub-tasks, each decomposed into a further series of sub-actions and/or sub-sub-tasks and so forth, i.e., *decomposition*. The structure dictates the underlying sequence of actions performed during the activity, but additional flow information can also be captured in a plan that specifies repetitions and choices.

The result of this process is illustrated in Figure 5.4. A task is first expressed in terms of an overall goal ("Sending a picture" in Figure 5.4) that the user is required to attain. Then it is re-described in subgoals ("1. Open the phone," "2. Activate camera," "3. Shoot a photo," and "4. Send it") and a plan (to send a picture, plan 1-2-3-4) governing when each subgoal should be carried out. Each subgoal is then reviewed to decide whether a solution is forthcoming or whether further re-description is needed. That is, an overall goal can be conveniently represented as a hierarchical description. This decompositional description has at its top an overall statement of the task in a verb-noun form, e.g., "Sending a picture." This overall description is then broken down, into more or less detail, by a process of progressive re-description.

The units of analysis are the verb-noun form, which is an instruction to carry out an activity. In Figure 5.4, the subtasks, such as "Hold the phone correctly" or "Press the shutter button," should be the relevant task components to the higher goal "Shoot a photo." Two notes are needed here. The first is that only a limited amount of information can be recorded in the boxes. This leads to a need for some other forms of supplementary information, typically a table or cross-references to the hierarchical diagram. A second feature to note is the presence of "procedures" or a "plan" on this diagram that represents information flow.

In understanding work, HTA is best seen as a general approach to analyzing tasks rather than as a tightly formulated methodology. As such, the key features are a hierarchical representation of a task, flexibility in the level of detail of information collected, and a tailoring of the analysis to the purpose at hand.

There have never been any strict rules on the details of the approach, other than the observance of some guiding principles. One of the guiding principles is that at each stage of re-description, a deliberate decision should be made on whether further breakdown is necessary. This decision on further breakdown is a key guiding principle of HTA. The level of detail of the analysis is determined by the nature of the task and the context in which it occurs. For instance, if the task is a quite difficult or relatively newly added one, it should be further broken down. In contrast, a simple task with no cost implications does not need further detail. As a second guiding principle, one problem that the approach has had is the misinterpretation of some of its features. The most serious example of this is when it is assumed that the hierarchy represents a model of cognitive activity as well as the details of the technology being used. That is not true, though the designer may have a certain level of imagination of a user's cognitive activities associated with performing subtasks. However there is no particular way to fully describe user's mental activities with HTA (except cognitive activities underlying the plan). It is thus the author's belief that the hierarchy in HTA should serve as a convenient form of representation of a complex entity, a task, which does not dictate to the designers a plain design suggestion that it ought to be designed in that way.

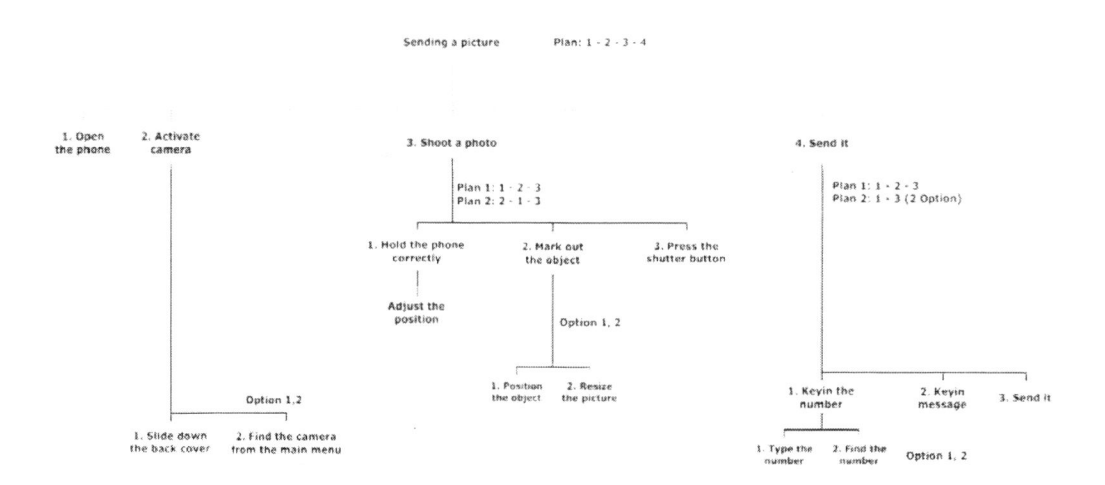

Figure 5.4. HTA of "Sending a picture" with Nokia™ N70.

Cognitive Work Analysis

The purpose of *cognitive work analysis* or *cognitive task analysis* (CTA) is to model the actions and especially the knowledge and thinking that users engage in when performing some task. HTA, described above, focuses only on the behaviours involved in task performance; CTA focuses more on the underlying knowledge, skills, and structures of task performance. The primary goal of CTA is thus to acquire a rich body of task knowledge about a domain and to assemble that knowledge into a user-centric work model.

Many CTA methods focus on describing users who manipulate a device in some way based on their understanding of the device, the procedures used to interact with the device, and the strategies required for solving problems associated with the device. These different forms of understanding make up the user's mental model of the task. For instance, TAG (Payne and Green, 1986)[37] provides a notation for comparing the consistency in task structures between interfaces. Interfaces that are consistent enable users to generalize actions based on prior knowledge or experience, which improve comprehension, retention, and positive knowledge transfer of those action sequences. In this way, CTA can rightfully elaborate the knowledge required to use the device. *GOMS (goals, operators, methods, and selection rules)* is one of the most prominent models in the HCI field, which does share the same underlying philosophy with HTA, describing very task-specific performance usually associated with some device.

Goals, Operators, Methods and Selection Rules (GOMS)

Card, Moran, and Newell (1983) proposed the goals, operators, methods and selection rules model, which is an engineering approximation of their *model human processor* (Newell, 1990). Whilst GOMS may not be a complete CTA, it may be implicative to allow design decisions to be made in terms of a natural mapping between the goal and the action, based on a user-centric mental work model.

---

[37] See Chapter 9 for further detail.

They postulated that users formulate goals (e.g., setting alarm) and subgoals (e.g., for this task, finding an appropriate menu item on the interface), each of which they achieve by using methods or procedure (e.g., moving the cursor to the desired location by following a sequence of keys). The operators are "elementary perceptual, motor, or cognitive acts, whose execution is necessary to change any aspect of the user's mental state or to affect the task environment" (Card et al., 1983, p. 144) (press the navigation key, move hand to the keypad, recall the appropriate menu name, verify that the menu item located is the right one, and so forth). The selection rules are the control structures for choosing among the several methods available for accomplishing a goal.

A very low-level version of GOMS is the *keystroke level model* (KLM; Bovair, Kieras, and Polson, 1990; Card, Moran, and Newell, 1980b), which attempts to predict performance time for users' performance of tasks by summing up the times for keystroking, pointing, thinking, and waiting for the system to respond.

Several criticisms of these cognitive work models – GOMS and KLM – have been made. Frequently, they assume an expert error-free performance without any interruption, interleaving of tasks or any consideration about interrelationships with concurrent tasks. And the underlying hierarchical goal decompositions may be too deterministic to reveal a user's mental work model to avoid errors. Some of the assumptions of these time performance models might be too extreme. For instance, some experts tend to employ short-cuts: they use the fewest number of actions possible to perform a task, and they can select task sequences from a repertoire of possible actions. This means that the choice of a particular action will be influenced by expertise and past experience; hence, the times taken or assumed can only function as approximations. Furthermore, the use of standard times cannot accommodate flexibility in performance strategy, nor can GOMS and KLM deal with time spent correcting errors. These entire problems stem from the fact that there is no mechanism for instantiating goals—they are just a given of the most likely user work model.

Interaction Unit Scenarios

To address the series of issues associated with either GOMS or KLM and other cognitive work analysis methods, alternatively, Ryu and Monk (in press) empirically demonstrated a finest and pragmatic approach – *interaction unit scenarios* – to user work modelling, by which an interface practitioner can instantiate how the user gets tasks done with a newly developing system.

The scenarios allow an interaction designer to make cognitively explicit both how user actions cause visible or noticeable changes in the state of the machine and how the user is expected to use this feedback to generate the next action. *Interaction unit (IU) scenarios* are constructed where each IU specifies one step in the cycle of interaction. Each IU specifies the visible system state that leads the user to take some (cognitive or physical) action. In addition, the IU makes explicit the state of the goal stack at the start and end of the unit and the mental processes (recall, recognition or affordance) required. In this way, one can describe the intimate connection between goal, action and the environment in user-machine interaction.

For instance, Monk and Wherton (*personal communications*) proposed a new stove design for patients with memory disorder. The key design decision in the new stove design (see Figure 5.5) was to model the correct actions and especially the knowledge and thinking that the patients would have when performing a task with the stove. In so doing, they created a new work practice, arranging the controls with salient flashing lights in the same configuration as the knobs, compared to the conventional stove design with merely the icons (no flashing lights, or none at all).

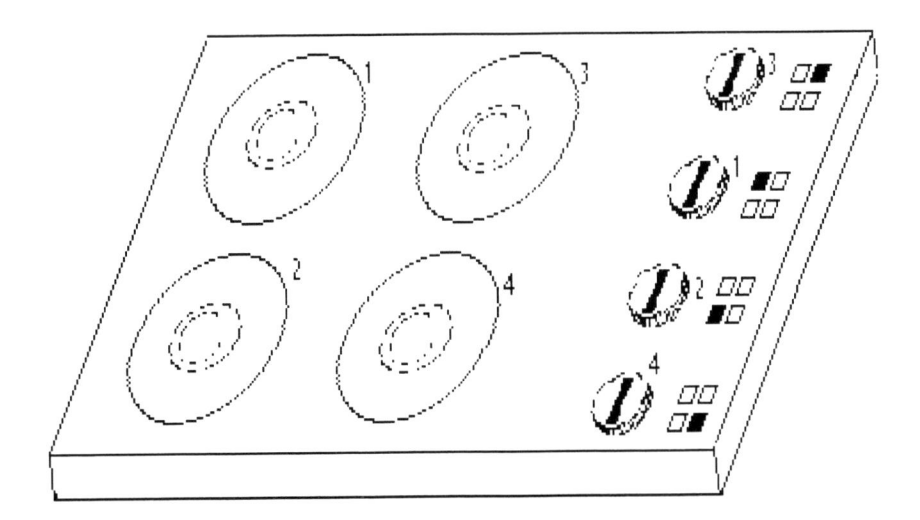

Figure 5.5. A stove design for dementia patients, example by Monk and Wherton.

As they suggested, the new work practice little had benefit from either HTA or GOMS to sketch out a user-centric work model. The hierarchical work models did not clearly simulate whether the new stove design would be well-fitted to the general cognitive capabilities of dementia patients. Instead, Monk and Wherton applied the IU scenarios to propose the new work practice as to how a dementia patient would get the task done with either the conventional or the newly developing stove, as shown in Figure 5.6 (conventional oven use scenario) and Figure 5.7 (new oven use scenario). As described in Figure 5.7, the IU scenario simulates a likely performance with the newly developing stove. The patient manipulates the new oven in some way based on their visual perception and understanding (i.e., affordance) of the information given (flashing lights), and the strategies required to solve the problem associated with the new stove. The new use scenario employs affordance thanks to the flashing light, which prompts the user to develop the correct action more easily. The flashing light on the new stove, as described in IU_2 of Figure 5.7, allows "Recognize p1 on or p1 with steady light" rather than "Recall p1 on" in IU_4 of Figure 5.6.

The IU scenario was inspired by the problems encountered by mobile system developers when designing the interactive behaviour of novel handheld devices, where the *de facto* standards that determine interaction design for systems with large screens, keyboards and mice are often inappropriate. Interaction unit scenarios allow the mobile interface designer to build up the interaction design in an iterative piecemeal fashion, testing by analysis each increment added. This would have a fine value in developing new work practices provoked by plain design suggestions in mobile interface design. See Chapter 9 for further detail.

Figure 5.6

	Environment			User Activity			Behavior
	Most Recent Changes	Other Information	Current Goal	Mental Process		Change to Current Goal	Action
				Recognition Recall Affordance			
IU₀	[START]		Switch on p1				
IU₁		k1, k2, k3, k4	Switch on p1	Recognise k1, k2, k3, k4		(-) Find knob for p1	Look at k3
IU₂	k3 focus of attention	k1, k2, k3, k4	Switch on p1; Find knob for p1	Recognise item not P1		(-) Find knob for p1; (+) Find another knob for p1	Look at k1
IU₃	k1 focus of attention	k1, k2, k3, k4	Switch on p1; Find another knob for p1	Recognise item is P1		(-) Find another knob for p1; (+) Turn k1	Turn k1
IU₄	p1 on: when Turn k1	k1, k2, k3, k4	Switch on p1; Turn k1	Recall p1 on		(-) Turn k1; (-) Switch on p1	END

Figure 5.6. A dementia patient using the conventional stove (i.e., no flashing lights). Her task is to switch on Plate 1.

Figure 5.7

	Environment			User Activity			Behavior
	Most Recent Changes	Other Information	Current Goal	Mental Process		Change to Current Goal	Action
				Recognition Recall Affordance			
IU₀			Switch on p with pan on it				
IU₁	[START] p/with flashing light and k1 with flashing light; when pan on p1	k1, k2, k3, k4	Switch on p1	Affordance turn k with flashing light --> p with flashing light On.		(+) Turn k1	Turn k1
IU₂	p1 on: when Turn k1, p1 with steady light	k1, k2, k3, k4	Switch on p1; Turn k1	Recognize p1 on OR p1 with steady light		Turn k1; Switch on p1	END

Figure 5.7. A dementia patient using the new stove (i.e., the flashing lights). Her task is to switch on Plate 1.

The Notation

This section describes procedures that serve as an effective springboard for the discussion and exploration of new interaction design with IU scenarios, by which a designer can make explicit the assumptions they are making when specifying how a user would interact with such as a new mobile system (i.e., proposing a new work model). These procedures include rules about how to detect unrealistic assumptions and so improve the user interface design. The IU scenarios approach is intended to make the cognitive processes of the user more explicit, i.e., the notation includes goals and cognitive processes associated with each action.

Figure 5.8 is an IU scenario for saving a document using Microsoft Word™[38]. Interaction units are to be read from left to right. An essential assumption behind the development of the IU scenario is that a cycle of activity starts with the state of the world and recent system responses; consequently, the environment is described before the other descriptions are represented. The first two columns specify the inspectable part of the system state relevant to the action specified in the last column, from *the perspective of the designer*. This can be thought of as how the designer relates their interaction design to a model of how the environment prompts a user to take some action. Inserted into this *designer's model* is a user component that specifies how the environment is related to the user's goals (i.e., presumably, *user's work model*). It is assumed that current goals condition processing of the environment, here *recognition, recall* or *affordance*, that leads to a new goal state which leads, in turn to the action.

Viewing Figure 5.8, as a whole, one can run down the last column to see the user's actions that form the scenario. The first column, "Most Recent Changes," describes the primary responses of the system to these actions. The other columns flesh out the implicit assumptions of the designer about how the user should be involved in this interaction in terms of goal changes and processing the display. The overall goal, "Save a doc on a floppy disk" in Microsoft Word™ is achieved by two actions: "Click" *MenuTab(File)* and "Click" *MenuItem(Save)*. The interaction unit scenario depicted in Figure 5.8 requires three interaction units to describe the task, the two actions (IU_1 and IU_2) and the last interaction (IU_3) to check whether the task has been completed or not.

Considering the Interaction Unit in detail, the "Environment" comes in two parts. "Most Recent Changes" describes the changes that result of the last action taken by the user.

For example, the most recent change "*Doc; MenuTab(File)*" in IU_1 of Figure 5.8 includes the initial information prompting the user, i.e., the current document and the "File" menu tab in the menu bar on the display; yet the most recent change "*MenuTab(File) Dropdown*" in IU_2 is generated by the last action "Click *MenuTab(File)*." Most interaction dialogues depend on a trajectory of system response leading to action, leading to system response and so on. In this way, interaction units are arranged in sequences to describe a scenario of use.

---

[38] For the reader's comprehension, the familiar example is applied rather than a mobile interaction design with the IU scenario that may vary from product to product.

	Environment			User Activity			Behavior
	Most Recent Changes	Other Information	Current Goal	Mental Process		Change to Current Goal	Action
				Recognition Recall Affordance			
IU$_0$			Saving a Doc onto a Floppy disk.				
IU$_1$	[START] *Doc:MenuTab(File).*	*Mouse.*	Saving a Doc onto a Floppy disk.	Recognize *Doc* is not saved; Recognize *Mouse* available; Affordance: Click *MenuTab(File)* --> *MenuFile* Dropdown.		(+) Reveal command.	Click *MenuTab(File).*
IU$_2$	*Menu(File)* Dropdown: when Click *MenuTab(File).*	*Doc; Mouse; MenuTab(File).*	Reveal command. Saving a Doc onto a Floppy disk.	Recognize *MenuFile(Save)*; Recognize *Mouse* available; Affordance: Click *MenuItem(Save)* --> *Doc* Saved.		(−) Reveal command.	Click *MenuItem(Save).*
IU$_3$	*Disk* Noise; *Menu(File)* Disappear; when Click *MenuItem(Save).*	*Doc; Mouse; MenuTab(File).*	Saving a Doc onto a Floppy disk.	Recognize saved *Doc*.		(−) **Saving a Doc onto a Floppy disk.**	[END]

Figure 5.8. A fragment of an IU scenario capturing the assumptions made by the designer of Microsoft™ Word of how a user should save a document onto a floppy disk in the PC environment. Items in italics are objects. Underlined words are changes in the display. An affordance is an expectation that some action will lead to (-->) some effect. Changes to current goals may be to add a goal (+) or remove (pop) a goal (−). The overall goal is in bold-face.

The "Other Information" column is also used to specify elements of the system that are not included in most recent changes but which the user must recognize as being pertinent to the task. For example, the "Other Information" of IU$_1$ of Figure 5.8 includes *Mouse* as the relevant object with which the user should interact. Designers may employ the "Other Information" column of interaction units to simulate and specify alternative scenarios. See Ryu and Monk (in press) for further detail.

The first column in the user activity part[39] of the IU is the current goal stack ("Current Goal"). This changes during the scenario as subgoals are added and eliminated. The goal changes that are assumed to occur in the IU are specified along with the processing resulting in these changes (recognition, recall, affordance). Take IU$_1$ in Figure 5.8. The initial goal stack is "Saving a doc onto a floppy disk," which is the overall goal. Recognition of the display causes a change to this goal stack, adding an intermediate goal, i.e., (+) Reveal command in the "Change to Current Goal" column. In IU$_2$ the new goal stack is thus "Saving a doc onto a floppy disk" and "Reveal command." In this way, the IU scenario allows developers to make explicit the assumptions about the users' mental processes that are inherent in their interaction design and to check how reasonable these assumptions are before testing with actual users.

Each action has separate explicit goals that must be reached to attain the overall goal of the task (Frese and Zapf, 1994). For instance, in Figure 5.8, the designers of Microsoft Word™ assumed that the action following from the goal stack in IU$_1$ and the affordance presented by the *MenuTab(File)* on the display is to "Click" *MenuTab(File)*. An individual user does not passively receive information, but actively selects information from the environment, formulating different actions and goals.

General instructions in describing IUs
1. Define the overall goal, e.g., "Saving a doc onto a floppy disk," in IU$_0$.
2. The overall goal is put in the current goal column in the first IU. Columns are to be read from left to right, in other words,

   The environment along with the current goal leads to recognition. e.g., both "*doc*" and "*Menutab(File)*" with a goal "saving a doc onto a floppy disk" lead to "affordance click *Menutab(File)* → *Menu(File) dropdown*." in IU$_1$.

   The environment along with the current goal and recognition/recall/affordance leads to the change to current goal, e.g., "*Menutab(File)*" with a goal "Saving a doc onto a floppy disk" and "Affordance Click *Menutab(File)*→*Menu(File) dropdown*" lead to the changes to current goal as "Reveal command" in IU$_1$.

   The environment along with the current goal, recognition/recall/affordance, and the change to current goal lead to the action, e.g., "*Menutab(File)*" with a goal "saving a doc onto a floppy disk" and "Affordance Click *Menutab(File)*→*Menu(File) dropdown*" and a newly generated goal, "Reveal command" lead to an action "Click *Menutab(File)*" in IU$_1$.

---

[39] The user activity part is comprised of the mental processes (changes to the goal stack and recognition, recall and affordance) and behavioural actions.

> 3. Add "[End]" in action for the last IU. e.g., "[End]" in IU$_3$.
>
> *Descriptions of environment and behavior columns*
> Step 1. Descriptions of IU$_1$
>   Step 1-1. Add "[Start]" in the most recent change column in the first IU, e.g., "[Start]."
>   As there is no most recent change in IU$_1$ the designer may specify what element of the standing system state the user will use to prompt the appropriate action.
>   Step 1-2. The other information column is also used to specify elements of the system the user must recognize that are not included in most recent changes, e.g., *mouse* as the relevant input device with which the user has to interact.
>   Step 1-3. Describe an action in the action column.
> Step 2. Descriptions of other IUs
>   Step 2-1. Include observable changes on the system by the last actions in the most recent change column. e.g., "*Menu(File)* dropdown" in IU$_2$.
>   Step 2-2. Include unchanged information on the system by the last actions in the other information column. e.g., "*doc*", "*mouse*", "*Menutab(File)*" in IU$_2$.
>   Step 2-3. Take into account user-initiated actions in each IU. Each IU can have only one action or set of actions that can be considered as one action. The action is some combination of interactions with the system that lead to observable changes in the system.
>   Step 2-4. Return to step 2-1 until no more action can be taken in the action column.
>
> *Descriptions of mental process column*
> Step 3. Clarify current goals in the current goal column in each IU, e.g., in IU$_1$, the current goal is the same with the overall goal. In IU$_2$, a goal "Reveal command" that is generated in IU$_1$ is put in the current goal column in IU$_2$.
> Step 4. Define the mental process in the recognition/recall/affordance column in each IU.
> Step 5. Specify the process of goal construction-elimination in the change to current goal column, e.g., the goal "Reveal command" is popped out in IU$_2$.
> Step 6. Return to step 3 until overall goal is completed.

Note. See Figure 5.8 and text for an explanation

Figure 5.9. General procedure of generating an IU scenario.

The IU model can then describe this active interpretation of the environment in the "Recognition/Recall/Affordance" column, which results in the description of how the goal stack has changed or what actions will follow. Here, an important note of the concept of *Affordance* is needed. Objects in the system have affordances (Djajadiningrat, Overbeeke, and Wensveen, 2002; Draper and Barton, 1993; Frese and Zapf, 1994; Norman, 1999), that directly guide users' expectations of what actions can be taken on the object (i.e., *physical affordance*) and what effects these actions will have (i.e., *semantic affordance*). For instance, in IU$_1$ of Figure 5.8, the affordance of the object "*MenuTab(File)*" is an expectation from the user that the desired effect will be accomplished, i.e., "*Menu(File) Dropdown*" (*semantic affordance*) when it is clicked (*physical affordance*). Of course, different users may have different constraints and expectations because they may have different experience and/or knowledge. Nevertheless, it is important to make explicit the affordances assumed when making choices between alternative designs; by doing so a designer can make better design decisions among alternatives. For further detail about affordance, see Chapter 6. By comparison, *recognition* refers to the understanding of the meaning of the object or environment. The affordance of an object strongly suggests the next action to be taken; however, recognition would be mainly used in reorganizing the goal structure. In IU$_2$ of Figure 5.8, recognition of "*Menu(File-Save)*" is used to remove the goal "Reveal command" previously generated in IU$_1$; by contrast, affordance of "*MenuItem(Save)*" strongly drives the next action – "Click *MenuItem(Save).*" *Recall* is the remembrance of a procedure or an event that occurred in the past. For instance, in IU$_2$ of Figure 5.6, a user can get rid of the overall goal "Switch P1 on," if and only if she or he remembers this command has been made. Such circumstance has the potential to be erroneous thanks to unnecessary cognitive workloads that lie at the heart of usability problems.

In developing new work models, the constructs "recognition," "recall" and "affordance" allow the designer to make explicit how the goal structure is changed in each IU and how actions are selected. They abstract over a lot of detail and cannot be counted as a thorough cognitive model. However, we believe that the interaction unit scenarios capture the essential assumptions in a mobile interaction design, which seem to be fair enough for mobile interface design practitioners as they are trying to propose new work models. Figure 5.9 provides some general steps for describing interaction scenarios with interaction units.

Putting IU Scenarios Together with Decomposition Processes

No work analysis technique can cover all design issues in order to propose new work practices, building upon the current work practices. IU scenarios work at a fine-granule level of abstraction of both system behaviour and the cognitive processing assumed of the user and thus add what we believe to be a designer friendly notation. Instead, HTA or GOMS can be treated in the same way to show how the designer moves through a cycle of considerations until the work analysis is complete.

The flowchart shown in Figure 5.10 captures the flow of decision making typically involved in our design process. Having combined HTA and IU scenarios, in HTA there is a concern where an accurate description of a cognitive operation is not possible. To resolve this concern, we propose a notation that describes cognitive and environmental features at the lowest level of description. Each interaction unit specifies one step in the cycle of interaction, that is: the visible system state that leads the user to take some action, the state of the goal stack at the start and end of the unit, and the mental processes (recall, recognition or affordance) required. This makes it possible to evaluate fragments of dialogue as they are designed by making explicit the assumptions of the designer about how the users get a task done; as a consequence, the designer can propose new work models to become acceptable.

The exploitation of the power of this combined approach is further described in Chapter 9 where IU scenarios detect some well-known problems in interaction design.

The IU model seems to be justifiable as it works well with decomposition in our putative interaction design as it is developed. However, a question is how the IU model will scale up to handle complex real designs. Building IU scenarios for all of the tasks selected requires effort from the designer. If the designer is developing an interface based on a well-established set of interaction methods (e.g., Windows™ CE platform), there may not be sufficient

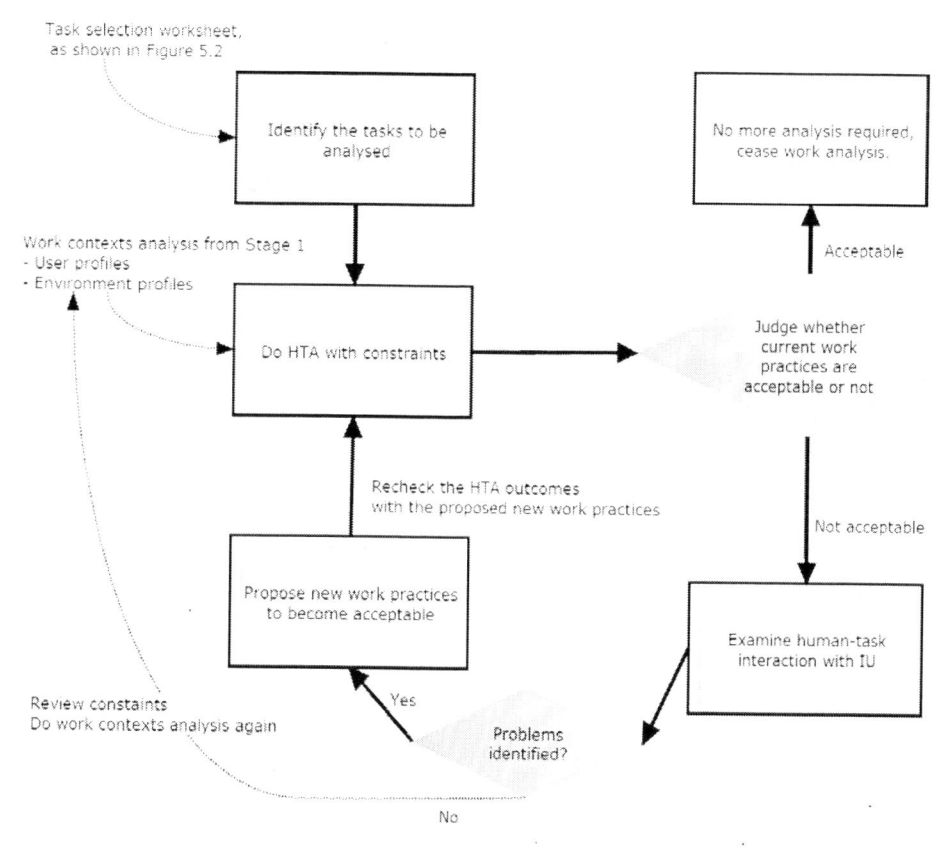

Figure 5.10. The cycle of work analysis. Working through this process enables the interface designer to effectively combine HTA and the IU model.

benefits to justify this cost. However, in mobile user interface design, the designer is often faced with inappropriateness of interaction techniques for the conventional user interfaces, which are not easy to extend to these devices[40]. As mentioned in Figure 5.10, as it is combined with decomposition-based work analysis technique, IU scenarios can model these more difficult design situations as fragments of interaction and there may be no great advantage to providing a complete set of scenarios for the design. We have as yet to gain experience of IU scenario analysis in real design projects (though I have successfully applied it for some small-scale mobile system developments) and so how this will work out is an empirical question. However, IU scenarios are presented here as a descriptive tool that captures the minimum detail required by designers of nascent technologies allows them to reason about a wide range of novel user interfaces, consequently being able to propose new work models from which we, as mobile interface design practitioners, would have a certain benefit.

---

[40] Revisit chapter 2 to refresh yourself about the different platforms that are currently being available to mobile interface design.

Chapter 6

TASK-FUNCTION MAPPING

In the previous chapter, we discussed the fact that work analysis (which is formative) can propose new work models. Each of the conceptual task components on which it is based must be closely linked with a particular class of systems design interventions in your prototype development. Here, this can be termed *task-function mapping*. This coupling process will ensure that the outcomes of the work analysis have direct, strong, and obvious implications for your interface design.

Yet, before we go further on task-function mapping, there are two issues to be raised in your awareness of your initial prototype design. First, this may be a good time for us to review what differences exist between mobile user interface and the conventional desktop computer user interface. Some of the design principles in mobile products may seem either unclear or less implicit because the domain seems to be relatively new rather than the traditional systems design approach. To address this issue, for instance, some PDA devices use Microsoft Windows Mobile™ which is basically a scaled-down version of Microsoft Windows™. Technically, this will cut down on the amount of time and effort people will have to spend on learning how to manage their files and applications on their PDA device. However, as you are well aware, we are often faced with the inappropriateness of the interaction techniques from a specific platform-based design style, which is not easy to extend to other mobile devices. Second, in many cases, we have already known the existing functions that are to be inherently included in the common mobile products, e.g., sending text messages in most mobile phones. Therefore, understanding the existing features would be another good starting point to develop a first prototype. Please note that Herbert Simon refers to this in his very broad definition of design as "everyone designs who devises courses of action aimed at changing existing situations into preferred one." This approach would guarantee congruence for the previously working products, and it will be in harmony with our putative design process. The design process we advocate here is well defined and highly structured, so that it can systematically harness what we need to consider in the new mobile user interface design. By including, for example task-function mapping as one of the essential ingredients of the putative design process, we could ensure both optimized features design with the existing products and comprehension of the new features included.

DESIGN CONSIDERATION: BRIDGING TASKS WITH FUNCTIONS

Simply transferring a full-sized computer interface to a mobile environment almost always results in a suboptimal mobile experience. Of course, this would be case by case, but it is generally said that desktop computer users are primarily interacting with the computer itself, but mobile users may primarily be interacting with the world, both through a mobile device and/or with other persons (Jones and Marsden, 2006). *Mobility* means reconsidering the entire purpose of the interface, not just changing display technologies or interaction styles from one to another.

The Physical Constraint: It Is Small

A small device, with a small display space, shows only a single window at any given time, with dialogue boxes and menus. As a consequence, task procedure design for mobile devices should be consonant with this constraint with a step-by-step task procedure design. The user can thus use exactly one set of display (information) at a time. The information cannot be accessed unless the user intentionally returns (or scrolls) his or her attention to the previous information display. The single window-based interaction generally creates challenges in accessing information outside the current display; for instance, the phone book may need to be seen during a voice call, as the person on the other end asks for the phone number of someone else. Because of this, effective information design would be the most important job of the mobile user interface designer, by which users could access any information resources that might be needed to successfully use the product[41].

The Carry Constraint: It Is Small Enough to Be Kept in the Pocket

Mobile devices are inherently different from computers in that they must be carried at any time. Otherwise, it is not the type of mobile device that we are currently discussing. This has several implications for the mobile user interface and service design. Certainly this ubiquity *per se* is associated with mobility as long as mobile devices can be viewed and used anywhere. This connectivity is maintained regardless of the location or movement of the mobile entity, but along with the accompanying characteristics, including:

- *Convergent device* – users will not carry a variety of single-purpose devices full time.
- *Personal device* – they are not shared with others, so it is very likely to be customized; the Japanese '*keitai*' is certainly not only a phone—it is also a personal item and part of the identity of many young teens in Japan.
- *Always on and always connected* – mobile phones are turned off only to preclude interruption for various temporary reasons.
- *Battery-powered.*

---

[41] Refer to Chapter 8 – Information Design – for further detail.

For such reasons, studies on the design of mobile phones are being conducted with the goal of trying to make them more wearable. For instance, Nokia™ has teamed up with a European fashion house to create clothes specifically designed to incorporate mobile communication technologies (Fortunati, 2002). The IBM™ pervasive wireless industry explores ways to make technology wearable and ubiquitous. NTT™ DoCoMo also announced the development of a new wearable mobile phone: a wrist phone. Samsung™ also announced its first GPRS watch phone in Europe at the end of 2003. These examples indicate attempts to make mobile communication more wearable, available and multi-functional without being uncomfortable. The issue raised in the mobile design trend is frequently an emphasis on aesthetics, leaving functionality in second place. For example, though internal antennas decrease reception quality, and smaller handsets have less durable batteries, the *carry principle* tends to cancel out the emergent functional features of the mobile user interface.

The Mental Constraint: It Is Small, So Affordance Is Essential

It is often said that an icon on a screen has an affordance to legitimize its placement there. For instance, users and designers have both learned that an abstract button on a screen will do certain things. In so doing, *affordance* is a term widely used in the HCI discipline, and most interface and product designs are often based on some notion of affordance, which counts in the IU model in Chapter 5.

Gibson's book, entitled *The Ecological Approach to Visual Perception* (1979), especially his theory of affordances, has been elaborated in interface design. Also, the book *The Psychology of Everyday Things* by Norman (1988) extended the meaning of affordances and systematically explored the roles of affordances, focusing functions, and natural *task-action mappings* in the design of interfaces. Norman used the term *perceived affordance* to mean the actual properties of an object (e.g., an icon, button and so forth) that can be perceived by the user. His definition focuses on how an object can be used by the user. His description also included cultural and social constraints, describing affordance as a phenomenon shaped by a person's previous knowledge and experience. When we say that an object has affordance, we mean to say that we have learned that it has a particular use or it works in a particular way. However, affordance becomes a very tricky concept for objects found in everyday use which become even more complex when we abstract them and then put them on the screen and, of course, user interfaces for many consumer products (e.g., mobile phones and PDAs) and Web sites in particular are expected to be more internationally and multi-culturally used.

Apart from describing one of the user's mental activities by the concept of affordance, it is of great value and importance in interface design as follows:

- It depicts a fundamental aspect of human cognition; that is, much information needed for perception and action is in the environment as invariants which can be picked up directly. That is, affordances are the allowable actions specified by the environment coupled with the properties of the organism. Therefore, to specify affordance in interface design, there is a need to describe it by either *constraints* or *allowable actions* in space, as exemplified in Figure 6.1.

- In most cases, constraints are the negations of allowable actions, but they further limit the allowable actions space. That is, the allowable actions are those satisfying the constraints' space and the constraints set the range of the allowable actions. If the affordances can be represented by these constraints, then the affordances are the disjunction of the constraints; otherwise, the affordances are the conjunction of the allowable actions, as exemplified in Figure 6.1.

Physical and semantic affordance of "OK" button

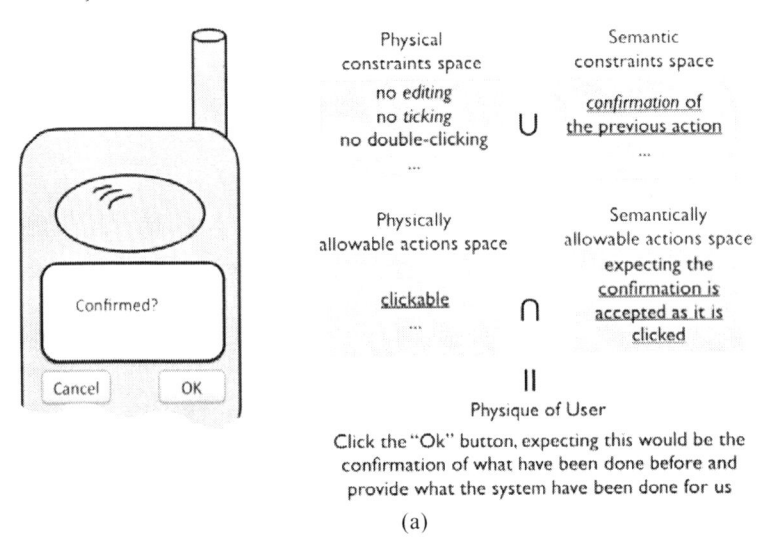

Physical constraints space
no *editing*
no *ticking*
no double-clicking
...

∪

Semantic constraints space
confirmation of the previous action
...

Physically allowable actions space
clickable
...

∩

Semantically allowable actions space
expecting the confirmation is accepted as it is clicked

‖

Physique of User

Click the "Ok" button, expecting this would be the confirmation of what have been done before and provide what the system have been done for us

(a)

Goal: Go back to "home", in light of first-time users

Physical constraints space
four icons at bottom
volume slider
two tabs at top...

∪

Semantic constraints space
playing *song list* has sometimes been used for on/off

Physically allowable actions space
click ▢ at bottom
touch ◁ at top

∩

Semantically allowable actions space

‖

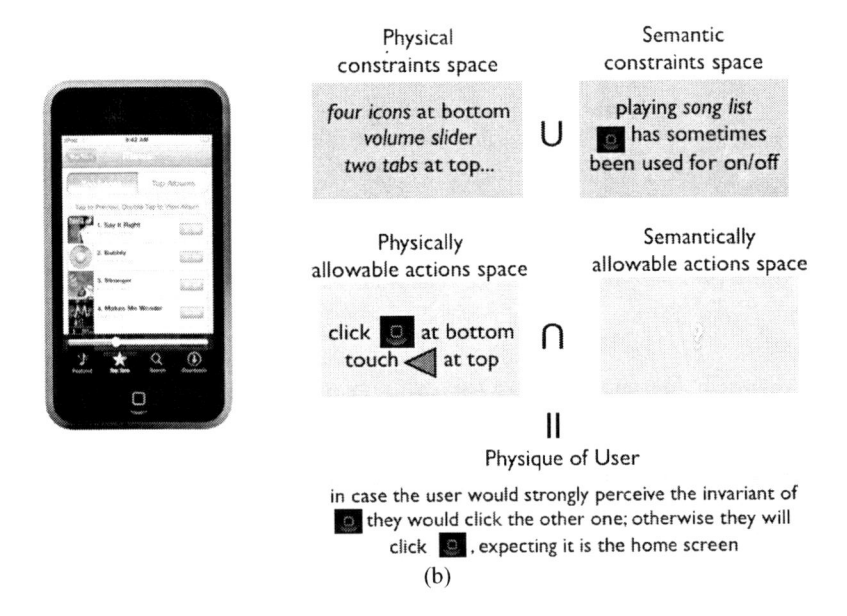

Physique of User

in case the user would strongly perceive the invariant of ▢ they would click the other one; otherwise they will click ▢, expecting it is the home screen

(b)

Figure 6.1. Affordance design in mobile user interface: (A) physical and semantic affordance of the "OK" button under the "setting" context; (b) physical and semantic affordance of the "Home" button.

Both physically and semantically, allowable actions' (and constraints') space indicate the fundamental properties of affordance design, as the complement between the environment and the user for the direct, effortless pickup of affordances. If you design an abstract button, for instance, in a mobile interface, and the user cannot match between what it is and how it can be used, the affordance of the button is being badly communicated. The key point to be noted here is the need for designers to look at earlier devices and see them as a useful frame of reference or source model of how forms and functions can be communicated via the concept of affordance in designing icons, controls and functions in your new mobile systems.

The New Trend: Direct Manipulation

Recently, one can see that conventional mobile user interface design (i.e., menu-based) has been replaced by relatively simple direct manipulations applied to visual representations of objects and actions, as depicted in Figure 6.2. In combination with affordance, *direct manipulation* has been the focus of attention in desktop computer interface design in the past two decades, but not until the recent success of Apple™ iPhone in mobile user interfaces.

It has great intuitive appeal and has been observed to provide performance and learning advantages over alternative interfaces. *Direct manipulation* interfaces have performance advantages because they closely resemble the "real-world" task environment that they are designed to support (Shneiderman and Plaisant, 2005). In so doing, they reduce the psychological distance between the system and the user, making it easier to perform tasks by requiring less adaptation on the part of the user (Norman, 1988). The emphasis is thus shifted to the visual display of user task objects and actions, in many mobile user interfaces.

In direct manipulation, it is much more important than before that the designer needs to create the metaphoric representations of interface objects and actions in relation to task objects, i.e., *task-object* and *interface-object mapping*. Users can learn interface objects and actions by seeing a demonstration, having an explanation of features or by trial-and-error.

Figure 6.2. Samsung™ haptic phone. (Downloaded from samsungmobile.com.)

In particular, the icon design needs to be *affordable* (implying the actions to be taken, and what effects they will present), *self-explanatory* and *transparent* (implying the interface actions visible to users), so that users can decompose their task objects into a series of intermediate interface objects. Once there is agreement on the task objects and interface objects (or actions), the ingredients of task-function mapping are ready for your first prototype.

FROM WORK ANALYSIS TO TASK-FUNCTION MAPPING

To make a mobile system work in practice, it is important to know how to map between the mobile work domain and its corresponding functional domain. For instance, for successfully using a mobile phone, users must construct a mental model as to what system function (e.g., "write messages with the keypad given, type the phone number and then send it") is designated to carry out a domain task (e.g., "sending a message to my friend"). These mental models are posited on a premise of the work analysis discussed in Chapter 5.

A good user interface helps the user quickly form a correct mental model for this kind of *mapping from tasks to functions*. Again, consider Apple™ iPod Touch. As shown in Figure 4.8, the mobile device clearly presents even a first-time user with the mental model about what the system is actually for. Eventually, in a first prototype design, the most important element to design is visualizing a mapping mechanism between task objects and their related interface objects on which people can perform the tasks with the functions given in your mobile system[42].

The first design exercise focuses on *congruence*, i.e., how logically system functions can be indicative of the work practices being proposed[43]. Put simply, we do not need to add up every function of your mobile system that users may like the least. Further, the functions in your system development must be justified by understanding the domain tasks that users want to accomplish with the system being developed. The work analysis methods discussed in the previous chapter thus submit themselves as necessary in this light. Otherwise, the *feature checklist* below, shown in Figure 6.3, would help you determine what functions should be embedded in your system development.

The feature checklist is a quick method to gather user information about the functions that would be better included in your mobile system from the user's perspective. It consists of a list of features of the system under investigation. The intention is that the user will recognize and remember these features as critical.

Please note that it is well known that users frequently tailor the design of their device and work practices to better meet their demands (see column 6 in Figure 6.3). In many cases, users make permanent changes that could have been, but were not, originally introduced by designers. This kind of permanent change could have been part of the original design if designers had the information required to introduce those features. This fact shows exactly why the coupling between proposed work practices and functions is important.

---

[42] There are many different styles of mapping mechanisms, such as direct manipulation, hierarchical menu-based, tab-based visualization and so forth.
[43] This further means to be equivalent in form between the user's tasks and system's functions.

	Environment				User Activity			Behavior
	Most Recent Changes	Other Information	Current Goal		Recognition Recall Affordance	Change to Current Goal	Action	
it₀	[START] Doc; MenuTab File;		Saving a Doc onto a Floppy disk.					
it₁		Mouse.	Saving a Doc onto a Floppy disk.		Recognize Doc is not saved; Recognize Mouse available, Affordance Click MenuTab File; → MenuFile Dropdown.	(+) Reveal command.	Click MenuTab File.	
it₂	MenuFile Dropdown; when Click MenuTab File;	Doc; Mouse; MenuTab File;	Reveal command; Saving a Doc onto a Floppy disk.		Recognize MenuFile Save; Recognize Mouse available, Affordance Click MenuItem Save; → Doc Saved.	(-) Reveal command.	Click MenuItem Save.	
it₃	Disk Noise; MenuFile Disappear; when Click MenuItem Save.	Doc; Mouse; MenuTab File.	Saving a Doc onto a Floppy disk.		Recognize saved Doc.	(-) Saving a Doc onto a Floppy disk.	END.	

Figure 6.3. Typical feature checklist question. Column 1: Existed? – "Did you know this function existed?"; Column 2: Used? – "Have you ever used this function?"; Column 3: How often? – "How often do you use this function?" Column 4: What for? – "Do you know what this function does?"; Column 5: Need? – "How often do you have any need for this function?"; Column 6: Personalized? – "Have you ever customized this function?"

Alternatively, in order to develop a mobile system that takes the needs of potential users into account in its range of functions, it is also helpful to prepare potential use scenarios, i.e., *scenario-based design* (Alexander and Maiden, 2004; Carroll, 2002, 2003; Carroll et al., 1991). With the aid of a speculative and detailed description of a probable usage concept, a list of the functions for a mobile system can be formulated in a fairly precise and piecemeal manner.

Taking the lead from the structural levels of system function design for a closer view of our interaction design framework, *goal-action matching* ought to be used to dictate mobile user interface design. Young's (1983) *task-action mapping model* characterizes this coupling process, using users' mental models of the task domain and bridging users' task components with the actions available in the system. It provides, therefore, a theoretical approach to task-function congruence (i.e., equivalence in form between a user's tasks and a system's functions), which depends on the cognitive representation of the user.

User Task Domain	User Action Domain
Operator	<operation> "+"
1st operand	<1st number>
2nd operand	<2nd number>
Evaluation	"="
Value calculated	Answer on display

Figure 6.4. Task-action mapping of the basic calculation task for a calculator – extended from Young (1983).

User Task Domain	User Interface Domain
Saving a document	"Save" command
Learning how to insert a picture in a document	"How do I insert a picture?" question in a help system
Deleting a file	Trash can icon
Shopping for books	"Books" tab on a Web site
Learning about all tasks or features related to memory	"Memory" topic in the table of contents for a help system
Changing your password or viewing your account history	"Account" link on a Web site
Sending a text message	"Write" icon first.
Voice call	Type the phone number first, and then click the "Call" button

Figure 6.5. External task-internal task (ETIT) model of some user tasks.

The example the author used is the performance of a particular calculation, such as the addition of 2 and 3 using a calculator. The user performance can be described in the arithmetical task domain as the binary operator "+" applied to the two operands 2 and 3. In the action domain of key-presses and readouts from the calculator, the same calculation can be described as the typing of the key sequences " 2 + 3 = " followed by the reading of the result from the display. When the calculator is being used for basic calculation (i.e., addition, subtraction, multiplication, and division), the mapping between domains is simple, as expressed in Figure 6.4. Otherwise, it can be said that user task components are not supported by interface action-domain, i.e., *task-function incongruence*.

Moran's *external task-internal task* model (ETIT: 1983) lies in the same track: tasks in the user's world must be realized in the embedded functions[44] of a system, extending the action domain of Young's *task–action mapping* model into the system domain.

Task Object	Action Object	Interface Representation by Direct Manipulation	Interface Representation by Menu-based Interaction
Message	(Send)		
	(Compose)		
Music	(Send)		
	(Play)		

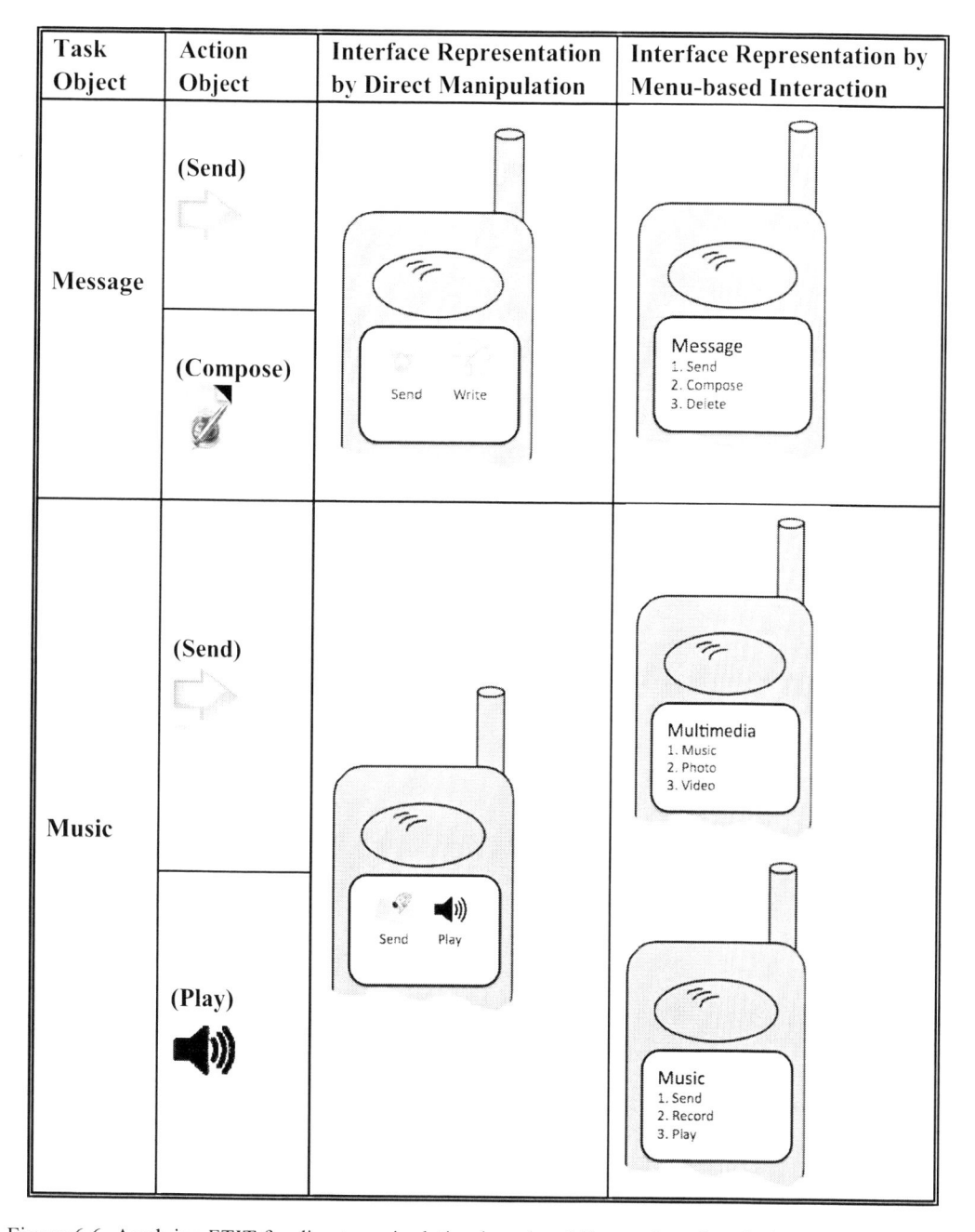

Figure 6.6. Applying ETIT for direct manipulation-based mobile user interface design.

---

[44] Note that Moran (1983) used the term "internal task" instead of "function" in his article.

In the ETIT model, the external task space means the set of possible tasks that the user can bring to the system (or simply users goals), and the set of possible tasks in the system is represented in the internal task (i.e., function) space, as depicted in Figure 6.5. By the ETIT framework, assume that the user translates each external task into a corresponding internal task (i.e., function), and this translation is expressed by a set of mapping rules.

Taking measures of the number of mapping rules is considered one of the criteria to judge the basic goodness of the user interface of the system given. Also, the ETIT model takes it for granted that work analysis scrutinizes the system being controlled to provide a great deal of insight into what functions are required for inclusion.

Inserted into Moran's ETIT model is *direct manipulation* (DM), which dominates one of the modern design interventions in mobile user interface design. Considering that its crucial feature is a continuous representation of user task objects and system feedback, the user task objects must be seen in the interface with their relevant visual displays to imply the results of an action before completing the action (*semantic affordance*).

Consider Figure 6.6. With the metaphoric representation (the first and second columns) of a user's task objects being available, designers can craft the interface objects and actions (the third column) to fit the user task objects. To the contrary, when the metaphoric representation cannot be well anchored to familiar task objects or actions, as a practical design guideline, the interface design can be tuned to a logical menu structure (as shown in the last column in Figure 6.6) to be relatively stable in the user's mental model.

Card Sorting for Task-Function Mapping

Apart from the *feature checklist* depicted in Figure 6.3, *card sorting* (Maurer and Warfel, 2008) is another information architecturing tool, which is quick, easy and inexpensive to carry out. It is used to help designers understand how users understand and think about organization and labels of functions in the system development. In a nutshell, it helps designers put functions where the users expect the information to be.

It normally generates an overall structure of functions in your interface design, and possible taxonomies. Also, it shows the mental models of the users, such as how similar those functions are and what those groups of functions should be called, and so forth.

There are two primary methods for performing card sorts:

- *Open card sorting*: participants are given cards showing content (or function items) with no pre-established groupings. They are asked to sort cards into groups that they feel are appropriate and then describe each group. Open card sorting is useful as input to information structures in new or existing interfaces.
- *Closed card sorting*: participants are given cards showing content (or function items) with an established initial set of primary groups. Participants are asked to place cards into these pre-established primary groups. Closed card sorting is useful when adding a new menu to an existing structure.

Analyzing card sorting is part science and part magic. Analysis can be done in two ways: by looking for broad patterns in the data (which I like best) or by cluster analysis software (statistical affinity tests would be another option for this purpose). When performing analysis on smaller numbers of cards[45], you may be able to see patterns by simply laying the groups out on a table, or taping them to a whiteboard.

Even though card sorting has many advantages, it does not deeply consider users' tasks, in the sense that card sorting itself is an inherently label-centric technique. If used without considering users' tasks, it may lead to an information structure that is not usable when users are attempting real tasks. A practical suggestion for card sorting is thus to combine with work analysis. Having identified all of the tasks involved in the current work practices, and then one can perform card sorting to structure functions in your interface design.

Card sorting is not a panacea to create an information structure. It is just one of them in a user-centred design process and should complement other activities such as, work analysis, and continual usability evaluation.

A Guide for Practitioners: Card Sorting
(adopted from Courage and Baxter [2005])

- Present each group with a set of cards. Each card should have written on it the name of a function that could be included in the new interface design.
- You may also include several blank cards to provide participants with the opportunity to add new functions (or menu items) of their own.
- Participants should then be told that they have to arrange the card into a potential structure for the new mobile user interface design. It is important at this stage to inform participants that there is no right or wrong way to go about this commitment. They should be told that all they have to do is to arrange the cards into a structure that makes sense to their group.

DESIGNING AN INITIAL PROTOTYPE

Prototypes serve as building blocks to both simulate design ideas and refine them, through iteration, towards a product specification that fits the requirements of the user population.

As shown in the sketches below (these were retrieved from a student project developing a new e-book reader), prototypes provide mobile system development teams with a common and effective communication medium. At the early stage, the prototype supports distillation and communication of the designer's understanding of the requirements. At the same time, the interface designer provides the user with an impression of the proposed mobile system. Prototypes allow testing at an early stage of development to avoid costly errors in the later specification. In this "conceptual and sketchy" phase, designers creatively try to come up with a range of design ideas, and are not yet actively trying to synthesize these into one complete and coherent design. Normally, for the first interface prototype, designers usually do not bother measuring the effectiveness, efficiency and user satisfaction of their design solutions. They merely want to get a feel for what design decisions to make, by implementing several functions on the prototype and observing users who struggle with their design ideas.

---

[45] Please do not submit every function to the card sorting method; instead, apply it for some unsure functions only.

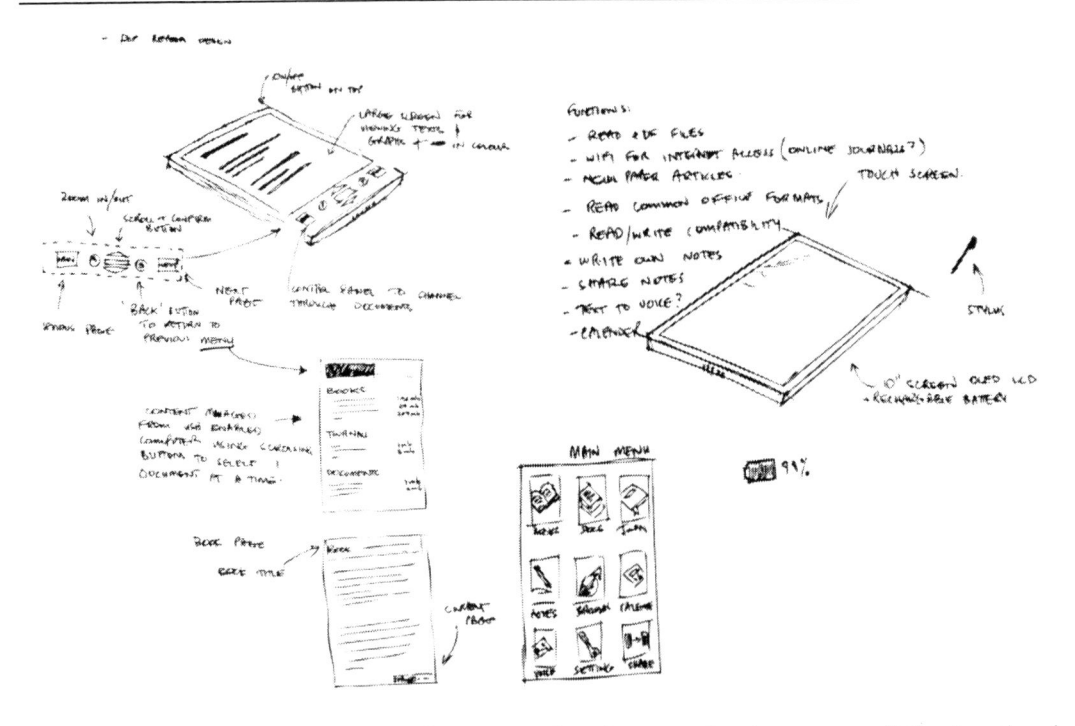

A prototype can take many forms, ranging from a sketch up to a fully functional simulation. The choice of prototyping approach depends on the product to be prototyped and the aim of the stage in the design process. Generally, in the first prototype design, the best are *low-fidelity* prototypes to have the end-users become part of the design process – *participatory design* (Ehn and Kyng, 1991). Participatory-design experiences are usually positive, pointing to many important contributions that would have been otherwise missed. In principle, more user involvement in the prototype development brings more accurate information about tasks, an opportunity for users to influence design decisions, and the potential for increased user acceptance of the final system. Hence, careful selection of users, presumably intended user groups, is important to build a successful participatory design experience. Participants may be asked to commit to several meetings, and should be told what to expect about their roles and their influence.

It is widely considered a good design exercise, in that user requirements tend to change over the course of development, overtaking initial design specification (Spence and Carey, 1991), so participatory design can present a communication tool in a tangible and accessible form to both the designer and the user. The process of prototype evolution incorporating user involvement allows user requirements to develop in parallel with the prototype (Beagley, Haslam, and Parsons, 1993). That is, the strength of participatory prototyping is its flexibility to evolve with the experience of both the designers and users.

Building upon the outcomes from Stage 1 and Stage 2 of the putative design process (refer to Figure 5.1), construction of a prototype requires the developer to make every design decision (or pending design changes), decomposing the requirements specification to a set of interface object components (Tanik and Yeh, 1989). Over the course of development, the prototype evolves from a loose concept into a solid simulation demonstrating the predicted benefits of the concept tool[46]. Applying such a functional prototype (i.e., a working prototype), using a representative population might allow the determination of recommendations of functional fitness of the system in an early stage of interface development. Of many prototyping techniques, most favourite is sticky-note prototyping.

A Guide for Practitioners: Some Guidelines to Shape Your First Prototype
(adopted from Jones and Marsden [2006])

- Design for truly direct manipulation. *Direct manipulation-based interface is successful not only because it makes the functionality more visible, reducing load on short-term memory, but also because its mode of operation fits well with people's desire to interact and manipulate the world in direct ways.*
- Design for ecological use.
- Design for maximum impact through minimum user effort. *Many usability guides encourage designers to keep things simple.*
- Design for personalization because mobiles are personal technologies.
- Design for play and fun. Users will appreciate the opportunity to have fun with their mobile. *To enhance the fun in your design, consider alluring metaphors, compelling content, attractive graphics, appealing animation and satisfying sound.*
- Design for one-handed use.

A Guide for Practitioners: Company's Pattern Libraries That
Help You Craft Your First Prototype

- Nokia's Symbian Design guidelines[47] explain the basic principles of visualization and graphic design, and provide examples of good and bad interface design. The document discusses graphic design for the mobile world—for example, colors, contrast, animation, and icon design—and gives tips on how to make the most of the small screen of a mobile device.
- Microsoft Windows CE platform also explores its style guidelines as being Web-accessible[48].

Sticky-Note Prototype in Participatory Design – A Mixture That Helps Mobile User Interface Design Become More Organized

Sticky-note prototype is based on the use of paper prototyping materials for mobile interface design that are specifically related to the physical size of the target platform (Parsons, Ryu, Lal, and Ford, 2005). The use of sticky-notes as a prototyping tool is just one part of a well-established practice of using tactile elements in the design of mobile interfaces. In this approach, the sticky notes serve two purposes. First, their size makes them ideal for the

---

[46] In actual fact, the very first interface awaits more substantial changes in the subsequent design processes, beginning from only a limited number of functions – often they are critical tasks – implemented. Also, quite often, the initial interface does not include any further consideration of menus, navigation or information design (see Chapter 8 for the prototype development encompassing information design). Even system feedback is not fully implemented (see Chapter 7 for the prototype development encompassing action-effect design).

[47] http://www.forum.nokia.com/info/sw.nokia.com/id/34762388-9434-4c42-9c5e-3e545b0975ea/S60_Platform_Visualization_and_Graphic_Design_Guideline_v1_0_en.pdf.html

[48] http://www.windowsfordevices.com/articles/AT4228820897.html

low fidelity prototype of the mobile interface screens themselves, but equally importantly their portability and ability to be arranged (and re-arranged) on a vertical surface makes them an ideal tool to model the mobile user's workflow, and to maintain the conceptual integrity of the task decomposition (Luchini, Quintana, and Soloway, 2004). Their key property is the ability to be moved into and between different environments and to link one piece of information to another in a precisely contextual way (Beato, 2005).

A number of different paper prototyping activities have been offered, and as overlays to other materials to show changes in content, such as showing pop-ups or items in a spatial environment (Liu and Khooshabeh, 2003). Another benefit of the sticky-note prototype originates from the fact that an early evaluation is possible, since the users can interact with the sticky-note prototype as though interacting with the real product.

In designing a mobile user interface, it is necessary to consider the various form factors of mobile devices in relation to physical constraints. Nehrling (2005) asserts that the specific advantage of the sticky-note medium in the prototyping of mobile device interfaces is that some of their standard sizes map readily onto the physical sizes of typical mobile devices. We have identified three generic categories of screen size: small (typical mobile phone windows), medium (Smart phones or Blackberry™) and large (PDAs such as the Pocket PC or Palm). The most common PDA screen is the Pocket PC format which has a screen of approximately 3 inches by 2.5 inches, a size which is available in some sticky-note ranges. To represent a generic mobile phone screen of approximately 1.5 by 1.5 inches, we can trim half an inch from a pad of 1.5 by 2 inches notes, as shown in Figure 6.7.

As well as using sticky-notes as screen proxies, we can also use different coloured notes when integrating different media objects (such as sounds or other things to be included in the interface) into the user workflow so that we could represent the various types of content and interactivity included in user interface design—say, red sticky-notes mean sounds and so forth, as illustrated in Figure 6.8.

This technical tip makes it possible to represent a total set of interface object components that act together but go beyond the visible boundaries of the sticky-note.

Figure 6.7. A sticky-note prototype for a generic mobile phone screen.

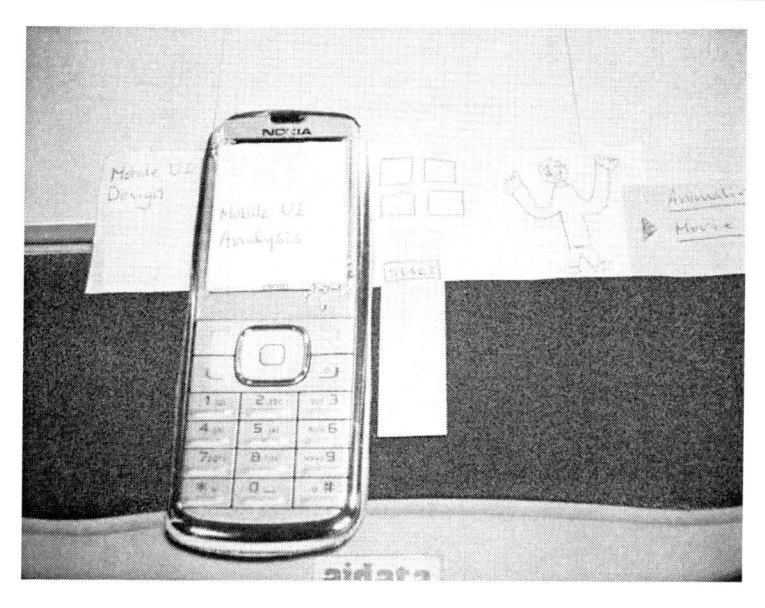

Figure 6.8. A prototype with different colors of sticky-notes.

For an exemplary color coding, the (standard) yellow sticker is for the main screen, with other chosen colors for categories of embedded or overlaying content, such as sound files, movie clips, applets or other media objects.

*A Guide for Practitioners—Prototyping Mobile Device with Sticky Notes
Can Apply to the Following Design Exercises:*

- Identifying needs and establishing requirements;
- Developing alternative designs that meet those requirements;
- Building a conceptual design;
- Navigation path determination;
- Accommodating for user error; *and*
- Prototype evaluation with participatory users.

Please note that sticky-note prototyping is highly recommended to be performed in combination with participatory design exercises (Monk, Wright, Haber, and Davenport, 1993). The ideal participatory design exercises with sticky-note prototypes have several iterations of a design-feedback loop, where the developers ask the users for their option as the implementation evolves. Collocation of the designers and users makes the iteration cycle tighter and more efficient. To boost this design-feedback loop, the Wizard-of-Oz technique can be effectively applied.

Wizard of Oz

Simulating sticky-note prototypes is essential to gain users' insights and implement various actions in a simple and efficient manner. In so doing, Wizard of Oz (Kelley, 1984) is a design concept assessment tool, allowing quick design assessment without detailed interface polishing (such as the size and colors of fonts). Normally, to assess an early design idea, two researchers are required: the normal researcher and the "wizard," who acts as the mobile system. The user points to a control and indicates the intended action, and the wizard adds or removes a sticky note, or even changes the whole screen in response.

A Guide for Practitioners Who Perform the Wizard of Oz Technique

- Encourage participants to interact with the buttons or widgets, not the screen directly. This will increase the fidelity of the prototype.
- Users should keep the feeling that they are in control.

In particular, mobile interfaces are quite likely to be multi-modal, and the Wizard of Oz technique can be an extremely effective method of assessing a multimodal mobile interface design; say, the "Wizard," in addition to changing what the screen does, also provides voice response and can even play recorded sounds for this purpose (Jones and Marsden, 2006).

PowerPoint Prototyping

In a very practical sense, I often employ Microsoft™ PowerPoint as a very first prototyping tool, as shown in Figure 6.9. Knight (2008) stated some good qualities of PowerPoint™ prototypes, including:

- *Everyone has it* – one of the great things about PowerPoint is that the people on your team usually have it. You can easily email a PowerPoint prototype to people for review and feedback.
- *Easy to make on-the-fly changes* – the PowerPoint mockup does not try to look exactly like the final product, so it is easy to work on high-level design issues and not get bogged down in details like colors or exact text.
- *Fast* – you can try something, hate it and try something else – all in a matter of minutes.
- *Everyone understands the semantics of a presentation.*
- *You can print it, beam it and collaboratively work on it.*
- *You can use it for click-throughs and animations.*

To begin with, create a simple PowerPoint mockup, each slide depicting a separate screen in your mobile interface (see the remote control in Figure 6.9). You can use shapes, text, and clip art to populate the screens. Once your basic mockup is in place, you can easily add elements such as animation effects on the controls and/or widgets.

Of course, PowerPoint is not right for every project. Here are some trade-offs to keep in mind if you are deciding whether PowerPoint is a good fit for your mobile interface design. For instance, imagine a menu structure where each menu entry leads to a separate details screen. You can do this in PowerPoint, but each individual page and each individual menu item must be created manually. It takes a huge effort. Please keep PowerPoint in mind for tentative interactions. As such, PowerPoint is great for testing interactivity, but it does not give you a realistic sense of what any one screen will really look like. Further, remember that PowerPoint screens do not scroll.

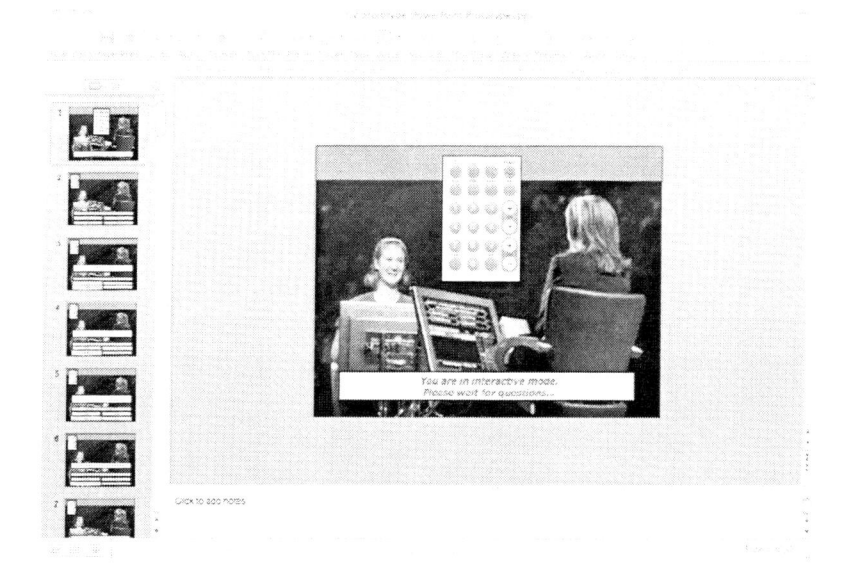

Figure 6.9. A PowerPoint prototype of an interactive TV interface design.

Overall, PowerPoint can be a blessing for mobile interaction designers who want to create interactive prototypes quickly and easily. Interactive PowerPoint mockups can give a flavour for how a mobile user interface will feel when the user moves through it—which is what interaction design is all about.

Chapter 7

ACTION-EFFECT DESIGN

In our first *task-to-function design* phase, discussed in the previous chapter, interface designers are creatively trying to come up with a range of design ideas, and those are not yet actively synthesized into one complete and coherent design. We observe that the *task-to-function* prototype usually does not bother measuring consistency, effectiveness, efficiency and user satisfaction of their design solutions. They merely want to get a feel of what design decisions to make, by observing users who struggled with their design ideas (i.e., the benefit of *participatory prototyping*). As many claim, the crucial part of the course of prototype development is that the prototype evolves from a loose concept into a solid one, demonstrating the predicted benefits of the concept tool.

In the next phase of prototype development, in this light, the first low-fidelity prototype of the mobile system is now needed to provide a strong foundation for the actual development of the mobile system. That is, task-to-function design needs to be evolved with elaboration of the functions gradually from a simple to a more detailed form with so-called *action-effect design*. Indeed, the first prototype contains a certain level of interaction behaviour with relevant actions and system feedback; however, they are usually very limited and not thoroughly synthesized into one consistent functional design. Hence, one rationale for our second prototype development is that the mobile user interface developers need the trial of pushing the inevitably incomplete first prototype through the arduous action-effect design process.

Again, simply applying desktop-based action-effect design practices for mobile interface design would not be good fit to create an optimal mobile user experience. In particular, modern mobile devices are normally covering many otherwise single products (e.g., MP3 player, camera, video-audio recorder, scheduler and so forth) into one, and knowledge transfer of the user's experience from one function to another is more important than ever before—for instance, taking a photo and recording a video seem to take on some level of affinity with each other. This *consistency*[49] of cross-functions in a convergent device has not been thoroughly investigated in the traditional action-effect design, so few paradigms are available when it comes to designing such a convergence platform, like mobile phones.

One way to make a convergent system appear simple is to make it uniform and consistent. Adhering to this principle – *consistency* – leads to a simple user's mental model too. Simple user models are easier to understand and work with than complicated ones.

---

[49] Another interesting definition of consistency is "memorable interactions from the past experience".

Design exercises in this chapter focus on this primary feature of interaction design which is directly affected by system implementation, i.e., the input (or action) or the output (or system feedback, system effect). Here, notably important is consistent action-effect design that crucially purports "ease-of-use" in mobile user interface design.

THEORETICAL FOUNDATIONS OF ACTION-EFFECT DESIGN

Perhaps the easiest way to design actions and system feedback might be to employ mobile user interface design guidelines for each platform (e.g., Microsoft™ Windows Mobile User Interface Design Guidelines[50] or Symbian™ User Interface Design Guidelines[51]). For instance, Rob Haitani's Palm OS design guidelines includes:

- Less is more.
- Avoid adding features.
- Strive for fewer action steps.
- Simplicity is better than complexity.

Such guidelines tend to be loosely organized collections of recommendations based on empirical outcomes, heuristics, and opinions as to designing system effects (such as dialogue boxes or menu design guidelines, and so forth). Unfortunately, the recommendations in typical guidelines are based on too few facts and experimental results and far too many opinions. Although well motivated, the soundness of the guidelines must be questioned on several grounds. First, even when supported by data, many of the recommendations made in guidelines are not valid for specific contexts. Another problem of guidelines is that they require interpretation, often by designers unfamiliar with psychological methods who are therefore ill-equipped for the job. Thus, the guideline approach to *action-to-effect* design may still result in inconsistently designed interfaces and actually increase the burden on interface designers by requiring them to consult large collections of ambiguous recommendations. In this respect, the rest of this section explores a theoretical account of action-to-effect design, by which the designer is able to lay out several guiding principles of action-to-effect design in mobile user interfaces.

Cyclic interaction theory (Monk, 1998, 1999; Ryu and Monk, 2004b, in press) is one of the basic accounts of human-information processing, stressing the intimate connections between action and system effect. Ryu and Monk (in press), very righteously, break down the interaction between the user and the system into three paths. Figure 7.1 delineates three paths in an interactive cycle: *goal-to-action* path, *action-to-effect* path, and *effect-to-goal* path.

Indeed, a common mobile phone use situation – setting a ring tone to "flight mode" – is straightforwardly explained by the three paths in cyclic interaction. We are able to simulate a possible action-effect design for the task along with relevant state descriptions as follows:

To perform this task, it is highly expected that the user naturally be given (by his or her intention) an overall goal (i.e., "*1. Setting a ring tone to flight mode*"). Next, she or he will

---

[50] http://www.windowsfordevices.com/articles/AT4228820897.html
[51] http://www.forum.nokia.com/info/sw.nokia.com/id/34762388-9434-4c42-9c5e-3e545b0975ea/S60_Platform_Visualization_and_Graphic_Design_Guideline_v1_0_en.pdf.html

seek to take a semantically relevant action or a series of actions with the interface objects on the mobile phone for accomplishing his or her current goal set on the *goal-to-action* path. It is highly reliant on the repertoire of actions in the specifications of the visible widgets on the mobile user interface, e.g., menu items to select, buttons to click or icons to touch and so on. In this case, it is most likely they will "*2. Scan the mobile phone, and then click (or touch, drag) a most semantically similar one.*" The action then triggers system effects on the *action-to-effect* path, i.e., "*3. A new screen appears (depending on system specification).*" Here, it is assumed that the designers already have the effects (or system feedback) in their system model or prototypes, by which the designers can simply work out whether the action performed at this action step would trigger predictable effects on the system from the user's perspective.

The system effects that do not make any sense to the user cannot create the appropriate next goals for the next correct interaction. Important here is thus what systems effects would be perceived as good fit to the subsequent interaction. Note that this chapter further discusses this issue later. In turn, when new system effects are presented, the following *effect-to-goal* path deals with changes in what is perceived by the users (i.e., "*4a. Search a menu-item, e.g., 'Ringing tone,' or similar*"; otherwise, "4b. If the system effect is not what they have expected, *Go back to the previous state and then try another scan*"), and continues to generate new goals, or eliminate completed goals in their interaction contexts. This newly organized goal set initiates another cycle, until they accomplish their original goal (i.e., "*Setting a ring tone to flight mode*"). Using these simple descriptions, an interface designer could briefly review their goal-to-action-to-effect design decisions (Monk, 1998). Note that the centre of this chapter is action-to-effect design, i.e., how to naturally drive the whole interaction given by the overall goal with an apt description of the action-to-effect path.

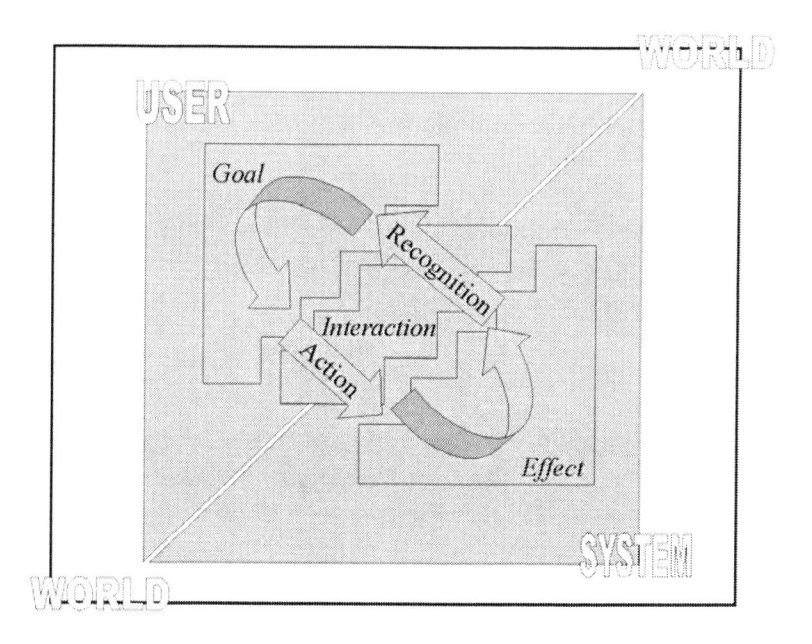

Figure 7.1. Cyclic interaction. It delineates a recognition-based cyclic interaction – extended from Ryu and Monk (in press).

The more thorough understanding that comes from the cyclic interaction theory may make it easier for designer to craft how to design actions and effects in mobile user interfaces. It is my experience that reveals that many action-to-effect designs have been worked out with the cyclic interaction theory, suggesting what actions should be visible at any time given and what systems effects would lead to the next interactions and so forth. Also, it is my personal belief that it is of great value because it helps the interaction designer to reason about what a cognitive process achieves, as well as what triggers that process; consequently, a better interaction design is prompted by action-to-effect design.

Also, designing interface objects (precisely, actions with interface objects) and system effects are heavily reliant on the concept of *affordance* (Djajadiningrat et al., 2002; Draper and Barton, 1993; Gibson, 1979), which was discussed in Chapter 6. Its most basic meaning is the effect a visible (or audible) object affords. Icons, earcons, photos or widgets on a mobile user interface are often intended to allow users to perceive affordances without learning, or at least not forgetting once learned. Another sense of affordance is whether users can perceive how to operate objects (or widgets) and what to accomplish with the objects, e.g., the button affords clicking and the label "OK" means confirmation. This affordance directly guides users' expectations of what action can be taken on the interface object and what effect(s) the action will have. Hence, any widget used differently from their own intrinsic affordance would result in usability problems.

A Guide for Practitioners: Two Features of Affordance

- The effects that a perceptible object intrinsically affords are mostly from visible objects (e.g., icons, labels, buttons), but more recent HCI techniques also allow other sensible objects, such as audible objects (e.g., earcons, specific key-click sounds), and tactile objects (e.g., drag and drop icons onto a specific region of a mobile user interface).
- It is the designer's interest that affordance—the effects that a perceptible object will generate, i.e., how to operate or activate the object and what this would generate in the system—is the defining substance for his or her system specification.

On the whole, both the cyclic interaction approach and the concept of affordance reveal what psychological information is required to design an action-to-effect path, and provide the designer with an ease-of-use theory to walk through in their design decisions.

Action Procedure Design

A system function is often sequentially or hierarchically organized in a sequence of actions. For action procedure design, it is worth returning to the outcomes of work analysis (see Chapter 5 for further detail). *Decompositional work analysis* is highly associated with naturally arranging a series of actions. Also, identifying a user's mental processes with IU scenarios would be substantially helpful to design in preference to both user-friendly action procedures.

Of course, a fundamental way to have the system as a whole appear simple is to make each of its parts simple. However, though you have designed each action simply, quite often the whole action procedure does not seem to be simple. For instance, in designing the mobile warehouse management system exemplified in both Chapters 4 and 5, a stocktaking task that currently uses six separate pages can be replaced by one single screen design with scrolls. The decision of an effective action sequence (e.g., scrolls vs. flips through a navigation button), in most cases, is not quite straightforward, and it all depends on the contexts and user preferences. For instance, the six separate page design would necessarily ask for more action steps with designated navigation buttons, but certainly does not require scrolling up and down and each screen page can be kept conceptually clean. In contrast, the one-page screen design with scrolls definitely does not suffer from the *cognitive momentum*[52] that comes from the dynamically organized page navigation (Freyd and Finke, 1984; Gamble, 1986; Thelen and Smith, 1999).

To clarify the account above, consider the chart wizard example in Microsoft Excel™. It helps users decompose the user task into a sequence of actions (i.e., four actions). In Figure 7.2, users first select "Chart Type" to generate a chart and select the area to refer to the data, and then specify the chart properties, and so on.

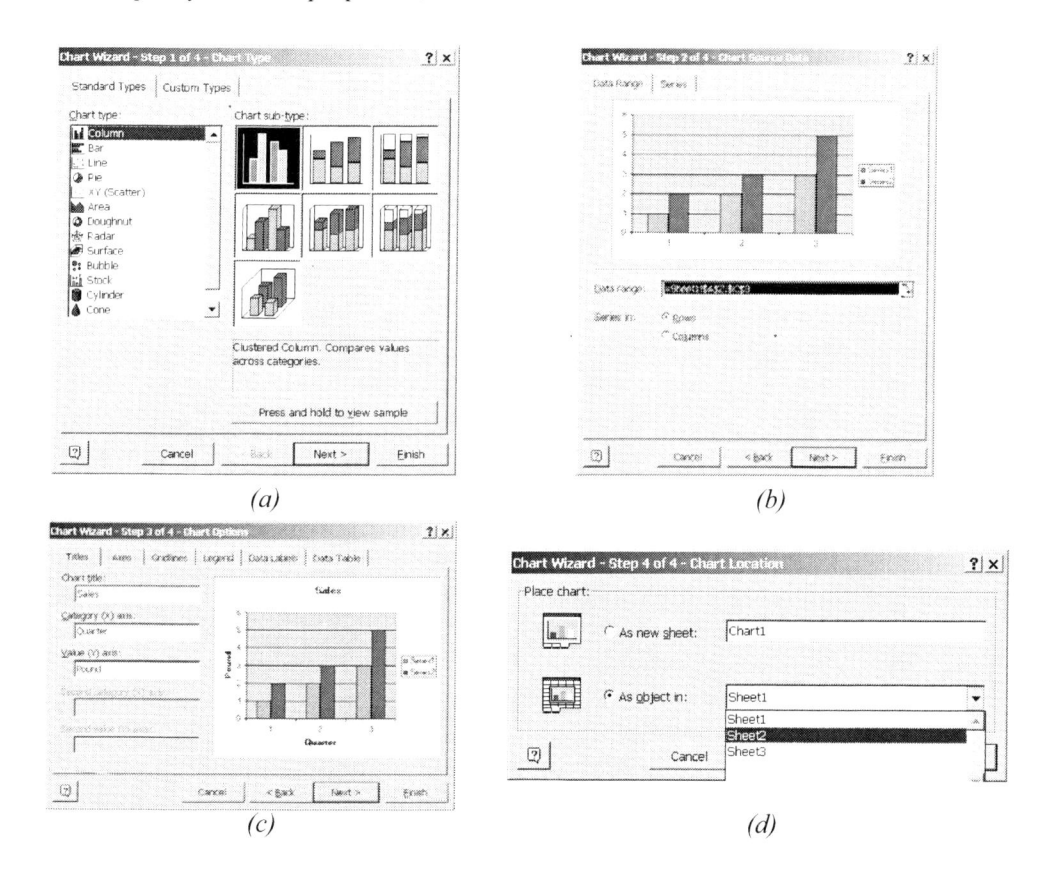

Figure 7.2. Chart wizard from Microsoft Excel 97. (a) Step 1/4 – select chart type, (b) Step 2/4 – select date to be referred, (c) Step 3/4 – select chart options, and (d) Step 4/4 – select chart location.

---

[52] Cognitive momentum fits with a picture of cognition as changes in activity over time. When a sequence of actions starts a trajectory of interacting internal processes, the trace of that trajectory is maintained for a period of time, giving us a perception of the future in the sense of building expectation (Thelen and Smith, 1999).

In this sequential way, they realize what has been achieved at each step and how many steps are left to accomplish the overall goal. The main benefit of the wizard application (or sequential action procedure) results from minimal information design in each action step. In contrast, making the four steps into one long action step with scrolls certainly creates a problem in locating the information that the user needs to find quickly, but will minimize the risk that arises from the wrong perception of the future event. Obviously, a simple action design is better than a complicated one if they have the same capabilities.

A Guide for Practitioners: Action Design

- Minimize redundant action procedures in a system. Having two or more ways to do something increases the complexity.
- Please bear in mind that the ideal system would have a minimum of powerful commands that provide all of the desired functionality and that do not overlap.
- More than four action steps should be avoided in mobile user interface design.

Mode Design

> A mode of an interactive system is a state of the user interface that lasts for a period of time, is not associated with any particular object, and has no role other than to place an interpretation on operator input. (Tesler, 1981)

Another important issue in action-to-effect design is *modes*. Many mobile user interfaces inevitably have *mode problems*. Modes can and do cause troubles by making habitual actions cause unexpected results. If you do not notice what mode the system is in, you may find yourself invoking a sequence of commands quite different from what you had intended.

To take an everyday example, consider a remote control that can be used to work both a TV and a video player. The same buttons are used to control each appliance; which appliance responds to a particular action depends on the mode the remote is in (the TV or video mode). Now, assume that this mode is changed by pressing a "TV" or "VCR" button on the remote control and that no clear mode signal provided. In such an imaginary case, if you want to play a videotape you must recall that the "VCR" button was pressed last. If someone else has used the remote control since you did, even this is impossible!

Yet it is seldom possible to remove all the modes from a mobile user interface, particularly when it is has limited input capabilities and small displays. Where modes are inevitable, the designer needs to review whether the user can easily see what the current mode is, i.e., what effect will result from the same action at a particular moment.

Monk (1986) argued that it is necessary to understand the user's cognitive representations to investigate the mode problem. That is, if the user is aware of what mode they are in, the inconsistent action–effect pattern is less of a problem. Therefore, it would be a mistake to conclude that modes are necessarily bad and should be avoided at any time in mobile user interfaces. More often than not, some modes can be helpful by simplifying the specification of extended commands. These modes work only for two cases in which they are visible and the allowable actions are constrained during modes. For instance, in 3G mobile phones, the actions for the camera must not be available for the other non-camera functions if, and only if, the current mode is clearly indicated by the system.

A *hidden mode* is the most extreme case that we must remove in every action-effect design, and it is relatively easy to detect because there are no external cues to mode. A weaker version of this is the *partially hidden mode* where the mode signal given has a relatively low salience. Finally, there is also another possibility of *misleading mode signals*. The latter two classes of "modedness" are not easy to track down; please refer to the key references (Ryu, 2003b, 2006; Ryu and Monk, 2005) for further detail.

FEEDBACK

With action design, a very core design decision by mobile interface designers is to provide users with as much information as necessary to enable them to complete tasks in an effective and efficient way. This includes appropriate feedback from the system so the user knows what is happening. In relation to this, one of the common problems associated with interacting with mobile devices is getting lost in a hierarchy of menus and sub-menus. As a close example of this issue, Brewster (1997) demonstrated that navigating information for phone-based interface (such as a phone-banking system) should be kept to a minimum because it can "get in the way" of the information or services that users are trying to access, and this could result in feelings of increased frustration. It is therefore important to design an instant feedback, in which each action by the user must be followed immediately by a response from the device. This response could be audible feedback or visual manipulation of the selected menu item or icon. Also, tactile feedback could be used to maintain an instant interaction dialogue between the user and the device, such as with a stylus pen or touch-sensitive interaction.

Metaphors and Icons

Mobile interfaces employ interaction styles that are different from those of the desktop computer environment. For instance, Pirhonen et al. (2002) developed a method of controlling a digital music player by using gesture and audio cues. Likewise, Samsung™ SCH-S310, a.k.a. *motion phone*, has a three-axis accelerometer and three-axis magnetic sensor that give it a range of motion-detection capabilities, and its fine trick is letting users write numbers in the air that the phone can then dial, as a remarkably new input method beyond the traditional key-in method adopted by most mobile user interfaces. The aim behind these products was to provide the user with non-visual feedback of interacting with the system, thus allowing them to pay full attention to their other physical environment. Also, like Apple iPod™ or iPhone™, there is a way of moving the finger across the screen from left to right (or the opposite), which would enable the user to start a track (or next photo), and moving it in the opposite direction would activate the previous track (or previous photo). The iPhone's metaphoric representation contributes greatly to the design's effectiveness, presenting congruent mappings between the real world and virtual world. For instance, (i) the movement used to search photos, i.e., moving the finger across the screen gives the same experience that one has with a photo album; and (ii) there is a high degree of consistent visual presentation—say, music or photo items densely displayed mean more photos or music in the folder.

Although good advances for designing new interaction styles for a mobile system have been achieved, the common interaction style being applied in most mobile user interfaces is still *icons* or *menus*. Icons can be designed to represent task objects and operations for the interface using meaningful interface objects and/or abstract symbols (see Figure 6.6).

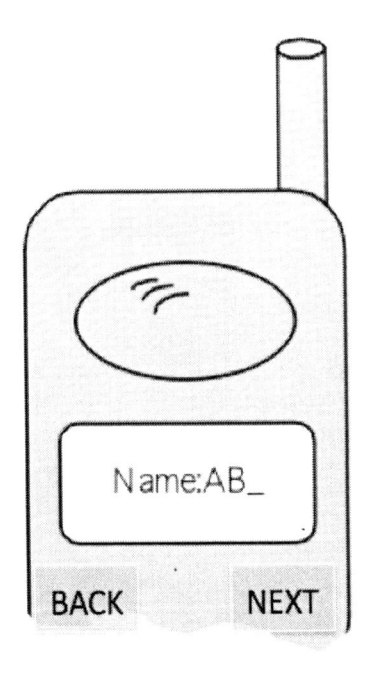

Figure 7.3. A congruent mapping between controls and their intended outcomes.

Icons presented in small device displays have much less screen real estate available, so they are typically designed to be simple, emphasizing the outline form of an object. Hence, the mapping between the representation and underlying referent should be *similar*, *analogical*, or *sensible association*. The most effective icons are generally those that are isomorphic, since they have direct mapping between what is being represented and how it is represented. Many operations in the mobile interface, however, are actions to be performed on interface objects, making it more difficult to represent them using direction mapping. Instead, an effective technique is to use a combination of objects that capture the salient part of an action through using analogy and association (Roger, 1989). For example, using a picture of an envelope to represent "message" provides sufficient clues for the messaging function in every mobile phone. However, if this problem is not fully considered, most icons would be more confusing rather than helpful, because interface icons look quite different and there is little consensus regarding a consistent one. Kim and Lee (2005) claimed that the same icon seemed to be differently interpreted by different cultural backgrounds, so that the generalization of the icons could be something that should be avoided in the design exercise. Again, when it comes to designing icons, the main objective should be designing a set of icons that are easy to recognize and are distinguishable from one another.

A Guide for Practitioners: Earcons

- Blattner et al. (1989) define earcons as being abstract musical tones that can be combined to produce sound messages to represent parts of an interface.
- In relation to mobile interface design, Leplatre and Brewster (2000) found that menus augmented with earcons produced better results in user performance than those found for the visual-only version of the interface.
- Many mobile phones employ a different-pitched sound corresponding to the key input.
- However, the impact of earcons has been limited due to background noise when using a mobile device in a public place.

Natural Mapping: The Position of Control Elements

In action-to-effect design, the functional assignment to interface elements – both physical and virtual – should logically gear towards the expectations of the user. For example, where elements are arranged horizontally, the reading direction from left to right should be taken into account[53]. This means that functions such as "Confirm" and "Next" should be on the right of the screen (because they semantically imply the next), and functions such as "Back" or "Previous" should be on the left, as shown in Figure 7.3. The vertical axis is semantically linked with the concept that "up" equals more and "down" equals less. This can be applied, for example, to functions such as volume and zoom controls. Figure 7.4 illustrates an unnatural mapping between the volume up/down properties and the ">>|" "|<<" controls. Yet, it is not true that this natural mapping is always possible. Jakob Nielsen's Alert box[54] sketched out this difficulty, building two different design rationales for the order of buttons in interface design.

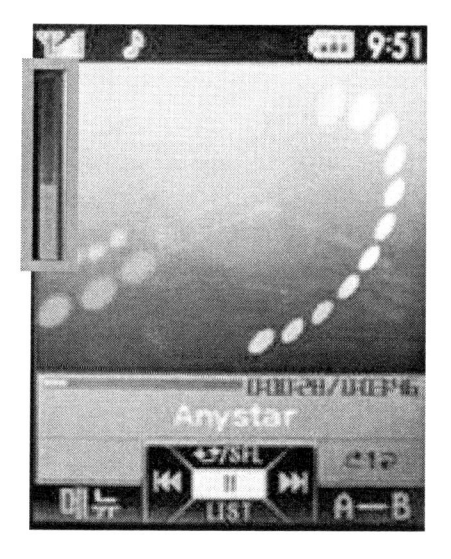

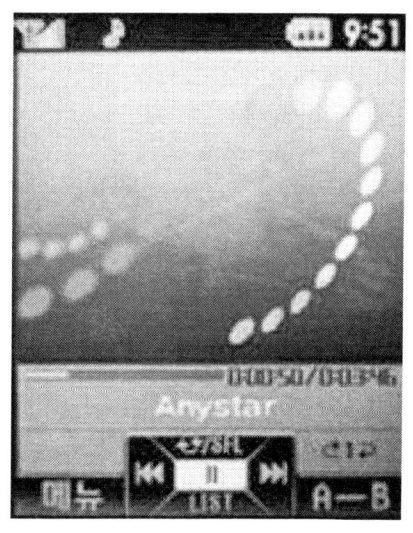

Figure 7.4. For volume control up and down, the action buttons ">>" and "<<" are to be used, which leads to very complicated and unnatural mapping rules.

---

[53] This is also true for the vertical layout of the reading direction from top to bottom.
[54] http://www.useit.com/alertbox/ok-cancel.html

See Figure 7.5. Considering the natural reading order in English and other languages that read left to right, you can locate the two buttons – "OK" and "Cancel" – as shown in Figure 7.5 (b) so that the reading order matches the logical order. Note that "OK" is a frequent action control.

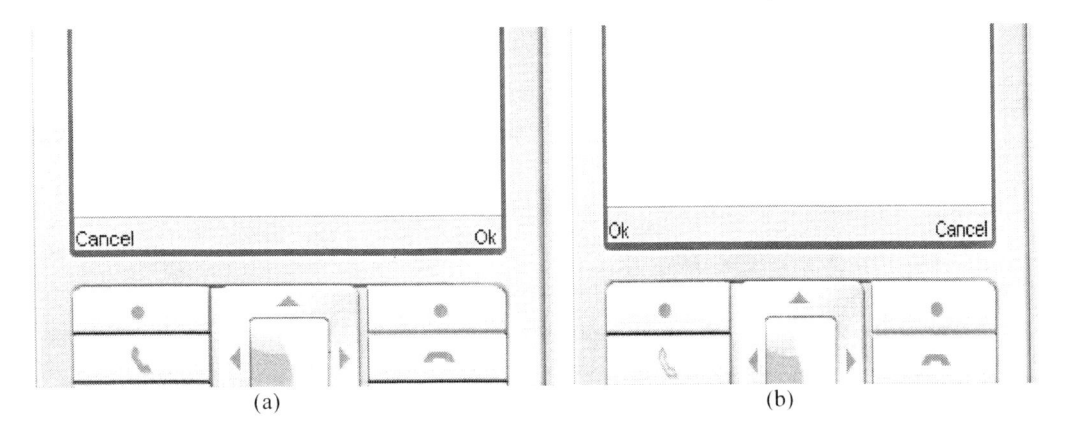

Figure 7.5. Where should the "OK" or "Cancel" button go?

On the other hand, considering the semantic flow, you could argue that the "OK" (as with "Next") is the choice that moves the user forward, whereas "Cancel" moves the user back. Thus "OK" should be in the same location as "Next" on the right. Nielsen suggested that this issue might be to some extent an endless and less useful debate, and to simply follow platform design standards or pattern libraries (for further detail on the platform standard, see the section "Action Design" below).

DESIGNING A SECOND PROTOTYPE

A new army of particular user interface styles (especially new ways for actions and feedback such as with icons, widgets, controls, dialogue boxes and keypads) is helping mobile user interface designers reach out to people to offer new mobile functions in a more practical way. However, there is no common style, due to manufacturer differentiation, manufacturer patents, and different needs with different capabilities. A simple low-feature scroll-and-select phone has been best offered. Nokia-style softkeys ("Options" and "Back" as the softkey labels) are common. Some phones have both an activation button ("OK") and softkeys. However, scroll-and-select phones with a large number of functions have suffered from the default tree hierarchy, in which the largest number of features force users to navigate deep into a complex hierarchy unless the desired feature is one of the small set that are readily assessable at the top level. Each user interface style has its own advantages and disadvantages, and this section on action-to-effect design thus focuses on these technical considerations, which include user interface style, input, and output modality.

Action Design

A number of physical input modalities have been developed for mobile interfaces, including a roller wheel positioned on the side of a mobile phone and a rocker dial positioned on the front of a device – both designed for rapid scrolling of menus and the up/down on the front face of the phone.

Recently, Apple™ iPhone has offered a novel input modality, in which users can use their fingertips to operate the user interface by tapping, flicking, and pinching to select and navigate (see Figure 7.6 for its gestural interaction styles). In an exciting way forward, there are actual advantages to using fingertips to operate a device. As one can easily guess, they are always available without a special input device, they are capable of many different movements, and they give users a sense of immediacy and connection to the device that is otherwise impossible to achieve with an external input device, such as a mouse. However, fingertips have one major disadvantage: they are much bigger than a mouse pointer and a stylus pen, so they can never be as precise as a mouse pointer. Additionally, there are some actions users can take with the combination of a mouse and keyboard that are difficult to replicate using fingertips alone, such as text selection, text cut, copy, and paste actions.

Physical interaction with small-screen devices reveals the conflicts of interest between creating the smallest physical size that will give the user mobility and flexibility while maintaining dimensions that are re-defined by the size and the motor functions of the human hands. There are two fundamental types of physical interaction: *one-handed* vs. *two-handed*. Some devices should only be operated with two hands, e.g., Nintendo™ DS. Mobile phones allow the user to dial a number with one hand, but interaction for more complex applications such as using the calendar, text message, or assessing the Internet is done with two hands.

Gesture	Action
Tap	To press or select a control or link (analogous to a single mouse click)
Double tap	To zoom in and centre a block of content or an image To zoom out (if already zoomed in)
Flick	To scroll or pan quickly
Drag	To move the viewport or pan
Pinch open	To zoom in
Pinch close	To zoom out
Two fingers scroll	To scroll up or down within a text area, an inline frame, or an element with overflow capability, depending on the direction of the movement

Figure 7.6. Apple iPhone™'s gestural inputs.

One-handed interaction presents the greatest challenges for the mobile user interface designer because the same hand is used for the interaction and to hold the device at the same time. This means that all fingers, except for the thumb, have restricted freedom of movement, and thus less scope for interaction. The key problem, however, is that the thumb has considerably less fine motor control than the other fingers. In one-handed interaction, a controlling system is located around the centre on the front of the device, so that the user can employ his or her thumbs to easily control the controlling system such as a mini joystick or a navigation button. In contrast, two-handed interaction often sees one of the hands playing a supporting role so that fine movements, such as data input through use of a stylus or keyboard, are possible. The concept is being looked at by many mobile phone manufacturers.

In spite of the promise of new input systems in a way forward, input with keypads is still common (this topic will be revisited later); but with the continuing evolution of mobile device capabilities, one sees that a device can use as input visual, auditory, or touch from the environment.

- *Buttons* – most are operated by pressing the physical buttons on the device or operating a stylus pen to press virtual buttons.
- *Speech* – this can be a very natural input mechanism, but difficult to implement, so speech input is only effective for short commands or data entries as of now. Speech introduces privacy and politeness issues in public spaces. It would be good for workspaces such as a personal office, where the system could interpret the input reliably. The intrinsic advantage of voice input is that this form of navigation cuts across the hierarchical levels of the overall deep structure, so each menu item can be called up immediately.
- *Stylus-pen* – stylus-driven devices have more flexibility in user interface, but some users reject stylus-pen use (e.g., because "blind typing"[55] is not possible) and find the hand-eye shifts between stylus and touch difficult to master and also suffer from parallax errors[56] (Tian, Kyte, and Messer, 2002).
- *Touch-sensitive* – devices such as Apple iPhone™ or Samsung's Haptic™ phone employ the touch-sensitive input system, with virtual buttons on the screen. However, note that the fingertips are much bigger than a mouse pointer, so they can never be as precise as other input devices such as a mouse pointer.

A Guide for Practitioners: Softkeys (extended from Jones and Marsden [2006])

Softkeys are dynamically assigned different functions, they should be positioned appropriately and close to the controlling system, and the current and active function of the key should always be displayed on the screen;

Consistency throughout the interface is important:
- Nokia –*the left softkey for "Options." All actions available to the currently highlighted item as well as the entire screen should be items within the Options menu; the right softkey for "Back," "Cancel," or "Quit" depending on context*
- Samsung – *the left softkey for "OK"; the right for "Menu"*

Ensure that the assignment of the physical keys is consistent; for example, the "cancel" functionally will always be assigned to the same physical key;

---

[55] Repetitive movements, performed without looking at a keyboard, are the base of blind typing. However, using a stylus pen forces the user to employ visual typing, so it cannot achieve higher speed and fewer misspellings, and leaves the user very tired.

[56] Errors from the observer's eye and pointer are not in a perpendicular line, which means that even when a user precisely taps a point that is believed to be correct, the point hit is generally several millimetres away from the one that he or she wants to select.

The link between the key and the screen should be also as consistent, close and unambiguous as possible.

Of the many design challenges in small mobile devices, text input can be regarded as the greatest. Slightly larger devices that can incorporate a touch screen with virtual keyboards and/or physically bigger key buttons are not so much helpful to work out. Cognitively, when using devices with small screens, *motor memory* becomes a particularly important factor, such as text typing in "blind" actions. That is why we still observe the standard mobile phone's conventional 12 keys: 10 for numbers 0–9, and the other two for the characters * and #[57]. Eight of the numerical keys also have three alphabetic characters each associated with them. This standard keypad design allows us to summon the motor memory from one mobile device to another.

A Guide for Practitioners: Softkey Design

In softkey design, traditionally, importance lies in how it can closely locate the softkeys on the display and their controlling keys on the keypad. In the recent mobile user interfaces, however, the softkey issue is minimal, because many of them employ touch-based input modalities.

By using the conventional 12 keys, a common text entry method–multi-tapping (for entering "C," for instance, you would tap the key "2" three times) is quite time-consuming and effortful. Silfverberg et al. (2000) have reported that compared with a conventional "QWERTY" keyboard, where people can achieve 60 or more words per minute (wpm), multi-taps rates amount to around 20 wpm for experts. The basic multi-tap method was significantly enhanced with the introduction of the dictionary-based predictive text methods patented by Tegic™, a.k.a. T9. Users press keys just once and the system, using a dictionary lookup, presents the most likely word(s) as the input progress. However, most letter prediction mechanisms create significant *cognitive dissonance*[58] if focusing on the letters. The user can be typing one word, but the screen is displaying another because that letter combination is more frequent. This can slow down the text entry process.

More recently, the *chording keyboard* has been explored as a possible text input method for small handheld devices, whereby characters are entered using combinations of key presses. Also, *ChordTap* (Wigdor and Balakrishnan, 2004) employs the within-group selection (e.g., for entering "C," for instance, tap the key "2ABC" for the letter selection and then the key "4GHI" to indicate "C" is the fourth letter on the key "2ABC"). *Twiddler keyboard*, adopted for mobile phone text entry (Lyons et al., 2003), is another alternative where multiple keys are pressed simultaneously to select a character on a keypad similar to that of a mobile phone. For example, to select the letter "m" the user would press the right key on the top row of the keyboard, and the middle key on the third row of the keyboard at the same time. Early results for both were quite promising, and showed that the techniques are significantly faster with fewer KSPC (keystrokes per character) than that of the multi-tap

---

[57] No one wants to re-learn the one-row keyboard, though its typing performance is better than the standard 3×4 keypad. For further detail on this, see Silfverberg (2003).

[58] An emotional state set up when two stimuli (i.e., the letter a user typed and the letter displayed) are inconsistent, or when there is a conflict between expectation and behavior.

method; however, we have yet seen any commercial success of these new types of text entry methods against the conventional multi-tap method.

Perhaps full-sized QWERTY keyboards may be more common in the near future; currently available are rollable fabric and infrared keyboards. Additionally, there are several variations of virtual keyboard. One such type is an input device that is available in the form of gloves, which interprets the movements of the fingers. Alternatively, a virtual keyboard (e.g., Siemens™ SX) can be displayed on the screen or projected onto an external surface. In either case through, interaction with virtual keyboards demands a greater degree of user attention and as such is unsuitable for mobile use scenarios.

A Guide for Practitioners: Minimize Text Entry in a Small Mobile Device (Ballard, 2007)

- Pick lists (drop-down menus or full-screen lists) *convert some tasks to cursor movement;*
- Global position system (GPS) or other location services *eliminate the need to enter current location for services.*
- Cameras *can take pictures of bar codes or other code systems.*
- Auto-completion *reduces keystrokes for long words.*
- Image recognition *of objects can be very useful.*

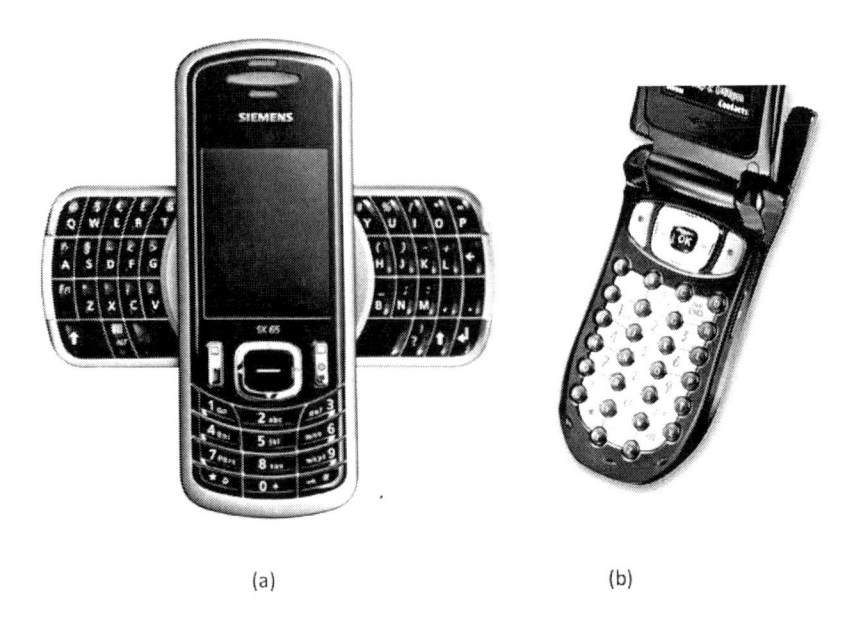

(a) (b)

Figure 7.7. Mobile keypad designs, extended from *Designing for Small Screens* (Dewsbery, 2005). (a) Siemens™ SK65 – a full QWERTY keypad; (b) Fastap (courtesy of Digital Wireless™).

Feedback Design

Audio Feedback

Audio displays can be played via the earpiece or the speaker. In many mobile situations, sound played via the speaker on the mobile device may not be a good option due to privacy or politeness issues. Sound played via the earpiece of course makes the user's ability to see and hear the display simultaneously more challenging. Therefore, the audio modality is still an appropriate technique for alerting contents such as emails or text messages.

Visual Feedback

It is common, and expected for all but sound feedback. However, key is that mobile devices are often used in a variety of environments and many of these will not always offer ideal or constant ambient lighting. Small screens prevent the user from smoothly reading large chunks of text. There are three reasons for this. First, it is easy to lose context when scrolling, as the physical and cognitive efforts of moving from page to page interfere with reading comprehension. Between-screen continuity is broken, due to the lack of *cognitive momentum* (Freyd and Finke, 1984; Gamble, 1986). Second, glare and pixel issues make the actual font difficult to read. Third, general text scanning behaviour (i.e., *saccadic movement*) is not supported. Most people scan text from nouns or phrases to comprehend text, but the frequent line and page breaks coupled with the lack of white space make this difficult to do, forcing users to read word by word (i.e., *pursuit movement*) rather than phrase by phrase.

To reduce the effect of the small display constraint a number of techniques have emerged to increase the display by virtual means. These techniques include zooming or panning, as well as the incorporation of windows and dialogue boxes. However, these are probably sub-optimal, so we will have to rethink the current convention of single-window screen design to multiple windows. This will be further discussed in the next chapter – Information Design.

Other Feedback

Tactile displays, such as vibrator, are accessible by some platforms. This is notorious for using up battery life quickly, but is important for getting the user's attention in a noisy environment.

Chapter 8

INFORMATION DESIGN

With the support of the *action-to-effect design* process, the prototype that has evolved from the first "conceptual and sketchy" one is now a somewhat high-fidelity working prototype. That is, all of the primary users' tasks are now well supported with this prototype, and the system feedback or critical information required for the user to perform appropriate interactions would exist at present with the prototype. Yet, the importance of effective task or information organization cannot be undermined in the final prototype, nor can it exaggerate the success of mobile user interface which, nowadays, is highly reliant on a visual and graphic manner of information organization.

User interface design history shows that designers have systematically designed and delivered information in an effort to share their perceptions of the system being proposed and persuade users to reach the same perceptions with them, i.e., to establish a *system image* through effective information design. As shown in Figure 8.1, for people to use a product successfully, they must have the same mental model (i.e., the user's model) as that of the designer (i.e., the designer's model). But the designer only talks to the user via the product itself (except book-length manuals), so the entire communication must take place through the system image (Norman, 1988, 1999, 2004)

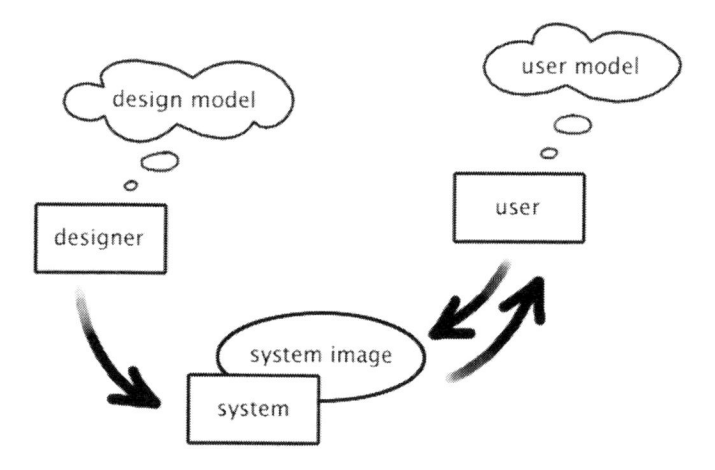

Figure 8.1. Information design and system image, adopted from Norman (1988).

This chapter reviews this information structuring issue in depth, to introduce our final prototype design, acknowledging the systematic and interactive nature of communication is the best to convey meaning and heighten understanding among all users involved, which are all about considering task-related organization, and organizing the information with graphical layout and selection mechanisms (Norman, 1991).

INFORMATION STRUCTURE

Best practices in the previous interface developments might introduce a good information design in the forthcoming interface design. Most information design in modern mobile devices is metaphoric in a way, or of its kind. Though some aural or textual information design is not impossible or ineffective, visual or iconic presentation makes it dramatically easier to use or comprehend and more aesthetic than other types of information design. Users can scan and recognize images or icons rapidly, and can detect subtle changes in size, color, shape, movement, or texture. They can even point to a very tiny area, and can drag one object to another to perform an action on mobile devices.

Often, when designers cannot produce appropriate direct manipulation-based interface objects, menu selection becomes an attractive alternative in average mobile interfaces. Whereas early mobile interfaces used full-screen menus with numbered items, modern mobile menus are usually pull-downs or pop-ups, all selectable by a tap of the stylus pen or keys on the front face of the mobile interface.

A Guide for Practitioners: Adjustments between the Platforms.

There are also a number of commercial browsers that automatically perform a sort of conversion to small screens. One of them is the Opera browser for mobiles. These sorts of adaptation systems work by identifying components in the content – e.g., menu bars or advertisements – and using some heuristics place them in a more usable sequence. This, of course, reduces the workload of the designer, but does not imply that we do not need to consider a new method of information design for mobile interfaces.

With the saturation of mobile technologies available and their competitive market circumstances, mobile information design tends to be quite patternized to be a good solution for standard mobile user interface design. While neither standard practice nor academic research has yet formalized what a pattern is and is not, patterns have become a good heuristic for a new mobile user interface design. Clearly there is no end to the list of all possible design patterns, and even the length of a book is not adequate for describing the majority of them.

There are two basic types of information design patterns: *UI designer patterns* and *corporate patterns*. UI designer patterns are a designer's solutions that likely work across a wide range of applications and on different platforms, but they cannot be coherent with a series of design exercises. In contrast, organizations with a complex set of mobile products may have their own set of highly specific and fully stylized "corporate design patterns." Many organizations, such as Nokia™ or Samsung™, standardize their design process using style guidelines and pattern libraries. Each pattern contains all of the same information as general patterns, with the addition of specific style requirements, a concrete visual design example, and frequently-used information design examples. Pattern libraries have several key advantages, such as guaranteeing consistency of user experience throughout their series of products. The patterns help the user interface become part of the brand along with the coherence visual design. This may be more important for mobile user interface design with the following advantages (Ballard, 2007):

- There is a significant reduction in the number of design decisions.
- There is higher compliance with device user interface paradigms across devices.
- There is reduced testing with regard to devices. A user interface built with patterns that were well-tested on devices is extremely likely to work on those same devices without failure or trouble.

In cases in which pattern libraries are not available, when it comes to structuring information on a small mobile device, it is worth revisiting the *gestalt laws* discussed in Chapter 4. They are truly helpful as a guideline for presenting information on small-screen mobile interfaces. As you know, we – human beings – tend to organize visual information into meaningful units according to a set of principles called the *gestalt principles of organization*. They are a series of rules that formulates the psychological perception characteristics of human beings. A designer can and should use these principles to organize information logically so that the user can understand content or information quickly and clearly. The following five laws summarize the most important principles of perception and how they affect the design of small-screen interfaces.

The *law of proximity* states that elements that are arranged closely together are perceived as a group or unit. This principle can be used to organize information and create units with a common meaning (like a sub-region of a Web page)—for instance, setting apart interface objects such as icons or creating units of meaning by displaying related items in the same color. Figure 8.2 shows an example that can be explained by this law. The four icons at the bottom, i.e., "Music," "Videos," "Photos," and "iTunes," as similar, follow the principle of proximity, so the finger can easily locate where to tap for playing media files without looking at the other icons. However, this design resource can only be used to a limited extent on a small screen due to the limited available space.

Designers can also let colors signal the similarity. This is often useful on the color screen, for instance, where two things must be placed apart on the screen but it is necessary for the user to perceive them as related (as exemplified in Figure 8.3). In this case, the items are simply given the same distinct color so that the user can see the relationship between the listbox and the button below the listbox in the same color. However, this only works where there are not a lot of other colors on the same screen.

A Guide for Practitioners: Color

Most modern mobile interfaces have a high-color screen. In particular, thanks to its aesthetic appeal, color-coding has become prevalent in many displays; however, note several limitations that may be critical for mobile system design.

Most important, population stereotypes (culturally and contextually, in particular) can produce poor design if a color-coding scheme does not provide a concrete meaning. For instance, in China, the highest color association in daily life is 70.4% (red with luck) and the lowest is 25.9% (blue and black with power), while the highest color association in the work context is 92.6% (red with danger) (Aykin, 2004).

Figure 8.2. The standing state of Apple™ iPod Touch. The law of proximity is applied. (Downloaded from Apple.com.)

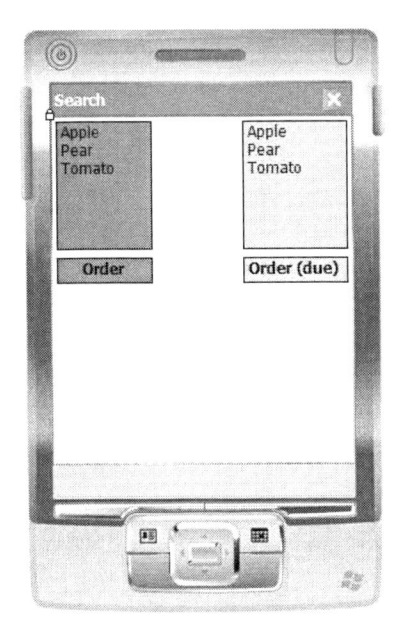

Figure 8.3. The same color coding for the items to be perceived as similar or linked each other.

The *law of closure* states that our perception skills will supplement incomplete elements in information design. This unconscious process can be used to create a visual link between the keys positioned on the mobile phone and the softkeys that are displayed on the screen. As depicted in Figure 8.4, "More" can be activated by the right button, and "Calls" for the left button. This autonomous mapping between the physical key and the softkey implies in practice that the *law of closure* can have a huge influence on a user's mental model of how the mobile system operates.

In mobile information design, one cannot undermine the power of *metaphor*. As we discussed in both Chapters 4 and 6, the human ability to learn is characterized by the creation of a link with what is already known, and the transfer of this knowledge to new situations. This principle is important, as it means that the designer must know the intended target audience for a product when selecting its information structure or organization, in order to ascertain what likely prior knowledge and experience the potential user can call on. The metaphor is thus a practical form of the mental model. The metaphor, which is reflective of human learning, starts with the familiar and transfers this to a new environment. For instance, the interface of the Pocket PC (i.e., Windows™ CE platform) is closely based on structures that are recognizable from larger Windows™-based PCs, and as such it can continue the connection with the mental model that is already familiar to the user.

Metaphors can be a disadvantage, however. The metaphor as a closed narrative framework has not become established; instead, a number of single metaphorical interface elements have become established, such as the button, the folder, the magnifying glass or the trash bin and, more or less, icons. In other cases, reportedly, there may be some confusion among the diverse user groups. In this regard, using icons in information design is useful and popular, but risky. For an icon to communicate successfully, the metaphorical image and its meaning in a digital context must be concretized first. To ensure this, redundant coding is used, i.e., the icon is supported by a textual explanation (i.e., labels), which the user can refer to if they are in doubt of the icon's meaning.

Figure 8.4. Softkeys and physical buttons closely located.

For most mobile applications, the top level of hierarchy and its corresponding tools available are represented only by icons. If the user then chooses to navigate deeper levels of the hierarchy, the information of the options is represented with more or less textual descriptions, in order to avoid the confusion by subtleties of the icons. Icons, therefore, are limited in their descriptive capability; once a certain degree of complexity is reached, icons become ambiguous and lose their advantages over textual descriptions.

A Guide for Practitioners: Spatial Memory and Icons

The arrangement of the icons on the screen should support a user's spatial memory capacity. This means that icons can be found quickly when the user can remember both the icon and its position on the screen.

MENU DESIGN

Whenever a limited set of possible actions is available, the potential exists for presenting the choices for display to the user. A menu is a listing of choices available in a particular interaction situation, involving a set of (1) actions, (2) objects of actions, or (3) description that qualifies either the actions or the objects, an arrangement of context-specific choices for user selection.

Menus are effective because they offer cues to *recognition* rather than *recall* of a command from a user's memory. Hence, users can indicate their choices with a couple of keystrokes and get feedback about what they have done. Simple menu selection is especially effective for mobile interfaces, users of which have little training or knowledge of what they have to use. However, a designer, in simply structuring menus for choosing from a pool of alternatives, does not guarantee that the interface will be appealing and easy to use.

There is little wonder that an average mobile interface has many more functions or tasks to be done. A traditional approach to form menus is sorting the functions or actions available by alphabetical order[59]. It enables a large number of elements to be made available in a sorted list. However, this approach does not convey a mental model that allows the user to accommodate all of the device utilities; this means that the user, with a menu structured by alphabetic order, cannot gain an impression of the information structure of the system. Very implicative to mobile user interface design is that several studies on the menu organization in computer systems (McDonald, Stone, and Liebelt, 1983; Shneiderman and Plaisant, 2005) demonstrated the superiority of a categorical menu organization over a pure alphabetical organization, particularly with larger menu structures. To create this categorical sort of mental model the organization rules must be plausible, categorical and the user must be able to rely on the organization rules once he or she has learned them.

On the whole, it is important that the menu structure creates a sensible, comprehensible, memorable and categorical organization relevant to the user's mental model. The information for organizing menu items can be obtained either from Stage 2, *work analysis*, i.e., hierarchical decompositions, or from *card sorting* (see Chapter 6) or *similarity analysis* (or

---

[59] Apart from alphabetical order, menu entries can be arranged in several different ways, such as ordered according to frequency of usage (probably high to low), ordered by category of entries based on some criterion meaningful to the user, or simply ordered according to user preference.

affinity analysis)[60]. However, classification of menu items is very personal, and culturally different. Even this changes over time (Park, Yoon, and Ryu, 2000).

A Guide for Practitioners: Menu Structure Design

- Categorization should be based on the outcomes of work analysis.
- Task-related structure should be the key design criterion of structuring a menu.
- If the number of actions available for a given screen exceeds ten, divide the list into frequent and infrequent commands.
- Provide numbered access to the frequent commands (as short-cut).
- If the device has an alphabetic keypad and the platform supports letter input, construct the menu with appropriate alphabetic shortcuts.

Different arrangements of menu items are also possible, either horizontally or both horizontally and vertically. As illustrated in Figure 8.5, individual menu items can be arranged horizontally in a bar menu at the bottom (Figure 8.5a) or at the top (Figure 8.5b). The lack of space on small screens allows us to arrange menu items collectively as well; examples are Nokia 6630™ using a carousel menu structure, and Sony-Ericsson™ W21S providing two methods of navigation – horizontal main menus and, at the same time, vertical sub-menus.

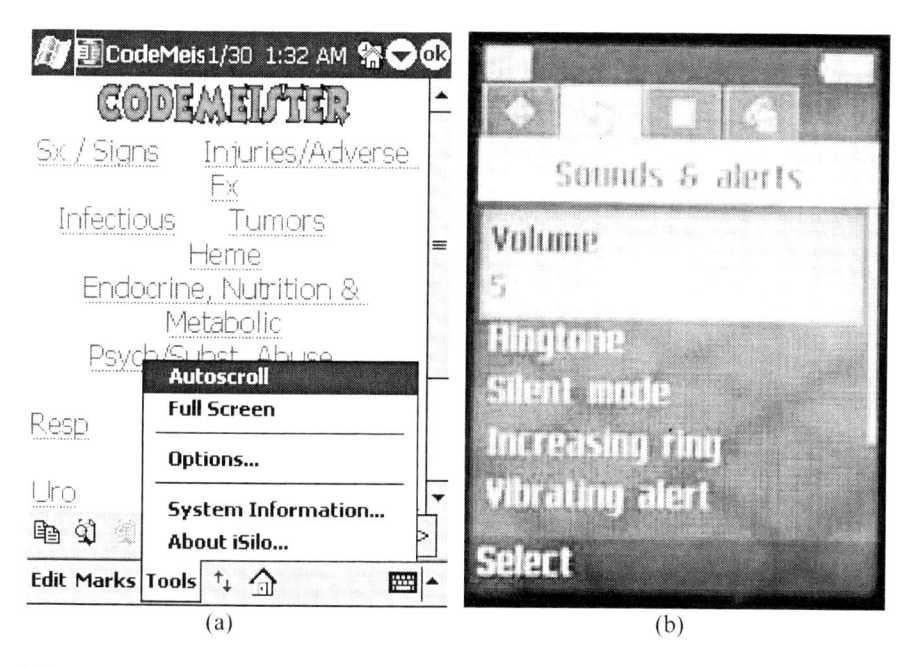

Figure 8.5. Arrangements of menu items. Pull-down and pop-up menus offer a selection of options that are presented in the list form. A pop-up menu opens up from the bottom edge of the screen (a), whereas a pull-down menu (b) drops down from the top edge of the screen.

Pull-down and *pop-up* menus have become established in many mobile applications on small screens. Ideally, pop-up menus are a better choice for devices where a stylus pen is used

---

[60] Please refer to Hastie, Tibshirani, & Friedman (2003), which provides a comprehensive and up-to-date

for the interaction because, unlike pull-downs, the menu items are not covered by your hand (Ballard, 2007).

A Guide for Practitioners: Spatial Memory and Menu Structure

In the case of Figure 8.5, to fully make use of human spatial memory, the basic functions must always be placed at, or pulled out from, the same and consistent point on the screen.

Various display techniques have also been offered to show several hierarchical levels simultaneously. For instance, a nested list is the one way to show the tree structure of a hierarchy; or, a preview of files in the form of thumbnails might correspond to a simultaneous display of the navigation and content levels. In the implementation of such concepts, small screens present a special challenge thanks to the limited available space. Indeed, a zoom function— precisely a distorted zoom—may help to make economical use of the space, but it can also be seen as less convenient in navigation. Here we discuss the two techniques: *list-based* and *table-based* layout. Figure 8.6 (a) shows the list-based menu layout, and (b) the table-based menu list.

Figure 8.6. List-based (a) vs. table-based (b) menu layout.

Mobile devices vary in their screen dimension ratios as well as size. For normal QVGA (240×320 pixels), most mobile phones are oriented vertically, with screens taller than they are wide. Hence, mobile user interface tends to be more often designed vertically, each link of which is on its own line, as in Figure 8.6 (a).

The table-based layout has generally a "standing-state" screen, with two or three columns of icons, from which major components can be started (see Figure 8.7 for Samsung™'s *Giorgio Armani* phone). The table-based layout screen seems to place the focus in the centre of the layout, not at the top. This can reduce the number of keypresses necessary to reach any given icon (Ballard, 2007). However, because a large number of menu items cannot fit on a single page, a table layout introduces extra complexity.

Figure 8.7. Samsung™'s Giorgio Armani phone (2008). (Downloaded from Samsungmobile.com.)

The user has to manage left and right cursor movement, up and down cursor movement, and page scrolling. This extra complexity can make the task of activating an item too complex. It can be widely said that the table-based layout is particularly effective for stylus pen and touch-sensitive devices.

A Guide for Practitioners: Designing Table-based Menu Layout

- Restrict the number of items to that will fit on a single screen;
- Reduce the icon size with clear labels;
- Present the most important items in the centre, so the user can quickly find and tap there.

Tab Design

Tabs, as shown in Figure 8.8, are another common mechanism used to arrange more controls that cannot fit on a single page. It is suitable for stylus-pen–based interaction or touch-sensitive screen. Tabs can be regarded as windows that are placed on top of (or beneath) one another, in a fashion similar to an index-card system. This organized principle is an efficient concept that enables users to call on many options quickly, using the combination of the tab and menu-based interface.

A Guide for Practitioners: Considerations in Tabs Design

- The categories shown on the tabs, which always remain visible (otherwise moded or greyed out), help to make the navigation of complex structures simple;
- There should be no more than five to seven tabs displayed at the same time (refer to Miller's Magic Number Seven plus/minus two[61]);
- Double rows of horizontal tabs should be avoided on small screens;
- Reduce the width of individual tabs so that more can fit within the available area;
- The horizontal layout imposes a limit on the amount of text that can be displayed, but this can be overcome by using icons.

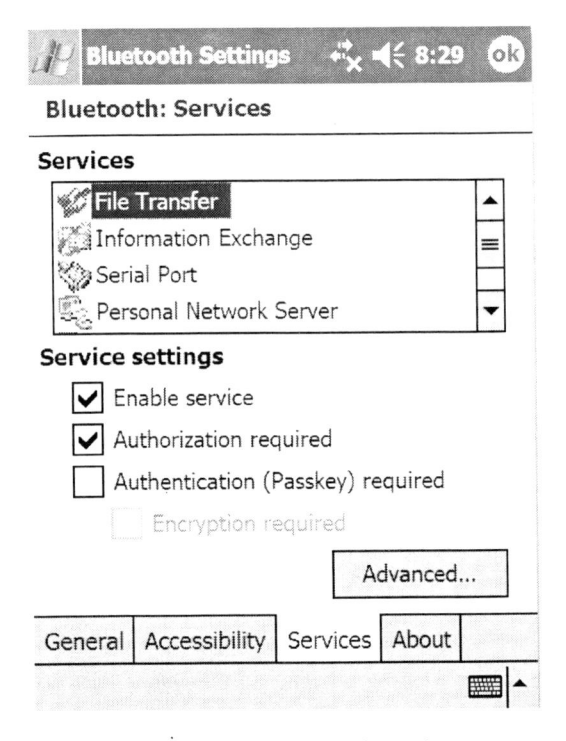

Figure 8.8. Tabs provide easy navigation in the complex information structure.

NAVIGATION DESIGN

Over and above organizing options with menus, interface designers of mobile devices often say that dealing with *navigation* is the most significant design decision and one of the trickiest areas in menu design. As discussed above, tabs in small mobile devices can logically make the menu navigation of complex structures much simpler, and save the current browsing session so that the user can easily return to it later. In effect, using tabs instead of a complex menu hierarchy reduces the user's cognitive workload.

---

[61] In 1956, a Harvard psychologist George Miller found that the short-term memory of different people varies, but found a strong case for being able to measure short-term memory in terms of chunks. A chunk can be a digit in part of a telephone phone number or a name or some other single unit of information. His research led him to discover a Magic Number Seven: most of the participants in his experiments were able to remember 7 ± 2 chunks of information in their short-term memory.

Although the tab-based menu design allows for easy access to a particular menu item under the single screen design, there are problems with this, such as dealing with many tabs at once. As discussed above, it is commonly said that there should not be more than five to seven tabs displayed at one time given, avoiding double rows of horizontal tabs on small screens. Dealing with multiple rows of tabs in one window especially worsens readability. Even finding a specific tab in a deeper level table-based interface (see Figure 8.6b) is always difficult for some people. Part of the issue with this difficulty lies in the lack of any sorting scheme. Tabs can be arranged without any sense of order, thus looking for a tab provides no meaningful understanding of a position of a tab relative to other tabs. Additionally, the clutter created by multiple tabs can create a screen design that is unusually small, with the tabs above or below it dominating the screen.

Thus, although tab-based menu design is adequate in environments where there is a minimal necessity for tabs (around five tabs or less), this scheme does not scale up, and alternate methods may be required to address this navigation issue. As such, the fundamental principle of navigation design is thus organization, not graphics. Although creative graphics can add to the aesthetic value of the navigation, the primary goal should be to make it easy for users to find their way to and from any part of the screen you design. One of the good examples of its kind is the "home" button on Apple™ iPod Touch or iPhone. It always leads the user to the "standing-state" (stand-by) of the system wherever they are, so that the user can easily start their navigation over from the standing state at any time.

However, please note that a good navigation design is a kind of craft, iteratively employing creative problem-solving skills to arrive at a practical solution by examining different alternatives. The most significant task that navigation design has to fulfill is to unambiguously guide users to the different sessions of the mobile user interface. Designers need to attempt to concretely explain which options are available and what the user should expect from the menu category (or item) once clicking on a corresponding label or icon.

In this light, work analysis plays another essential role in exploring the alternatives. To make the perception of menu navigation easier, the navigation is often structured by the outcomes of work analysis to create prominent goal-oriented headings (or categories) with information mapping techniques (such as *card sorting* and *affinity analysis*). As a consequence, the structure of the menu system would resemble the structure inherent in the relationships among user goals or tasks; if so, users can easily go back when they are lost in the information space.

To communicate navigation options in a more effective way, designers often produce a set of appealing icons. In such cases it is also important to make sure that the icon is easily recognizable, clearly conveys the message, corresponds to the meaning it stands for and is not too small. Attractive icons are, of course, always preferred to boring ones.

A Guide for Practitioners: Navigation Design

- Avoid deep information hierarchies;
- Each bottom-level item in the menu organization, most likely, is somehow a representation of a user goal, and category labels seem to represent higher-level (or abstract) goals or categories of related goals;
- Keep all critical information within three clicks;

- Use positive guidance (Dewar, 1993), prompting appropriate behaviour of the user and their expectation at the time of interaction;
- Choose labels that are predictive, distinguishable, and short;
- Make text easily legible;
- Give users a constant sense of where they are;
- Use a consistent navigation device (such as a navigation bar) with elements that change to give feedback regarding the current location;
- Make it easy to backtrack and recover from errors;
- Allow users to change direction without backing up to another screen

On the whole, central to mobile navigation design is a navigation system structured in a way that allows a user to access a specific goal as quickly as possible. And the guidelines above may help you foster effective navigation design.

DESIGNING THE TEST-DRIVEN PROTOTYPE

For the final prototype design, *test-driven development* is now brought to the forefront of the interest. Test-driven development is a novel software development practice and part of the *extreme programming* paradigm (Beck, 1999). It is based on the principle that prototypes should be tested for a small coherent module iteratively, while the module is created. This is the opposite of what is usual in current prototype development methods in which testing of a prototype is often an after-thought, rather than a primary driving force of design. Once applied systematically and continuously, test-driven development is supposed to incorporate users requirement changes more easily, lead to superior technical solutions in interface design, and result in a better and cleaner interface prototype. In effect, a test-driven prototype is expected to motivate high-quality interface designs and systems, and drastically reduce design flaws in the final system implementation. For the nature of modularity in test-driven development, the test-driven prototype created depends on the questions that need to be answered by the demonstration. For instance, if the "mode" of interaction is to be a great concern, then the physical input elements and the system's corresponding feedback should be implemented as realistically as possible.

Various tools for test-driven prototype development have been offered. Simulation is one kind and, of course, the most popular with mobile interface designers. It can be programmed with a number of authoring programs, such as Macromedia™ Director, Adobe™ Device Central C3, Microsoft™ Visual Studio and so forth. In particular, Openwave™ allows you to easily develop various types of mobile phone UI designs, which is freely available (see http://developer.openwave.com). Many of the authoring tools above also include interface design style templates for various mobile devices (see Figure 8.10).

Recently, Apple™ iPhone Interface Builder (see Figure 8.9) has been introduced to help developers create some applications for the two new mobile devices, i.e., the iPhone or iPod Touch. It includes a simulator and an integrated development environment, so the designers can implement and test their mobile application as practically as possible. Like many other mobile application authoring tools, it also provides a complete and integrated process for developing, debugging, and distributing the applications for the iPhone and iPod Touch. More important, it presents an additional list, the Apple™ iPhone *corporate design pattern library*, which has information the developer can follow regarding the Apple design philosophy, and thus avoid a higher level of variations in their individual design exercises.

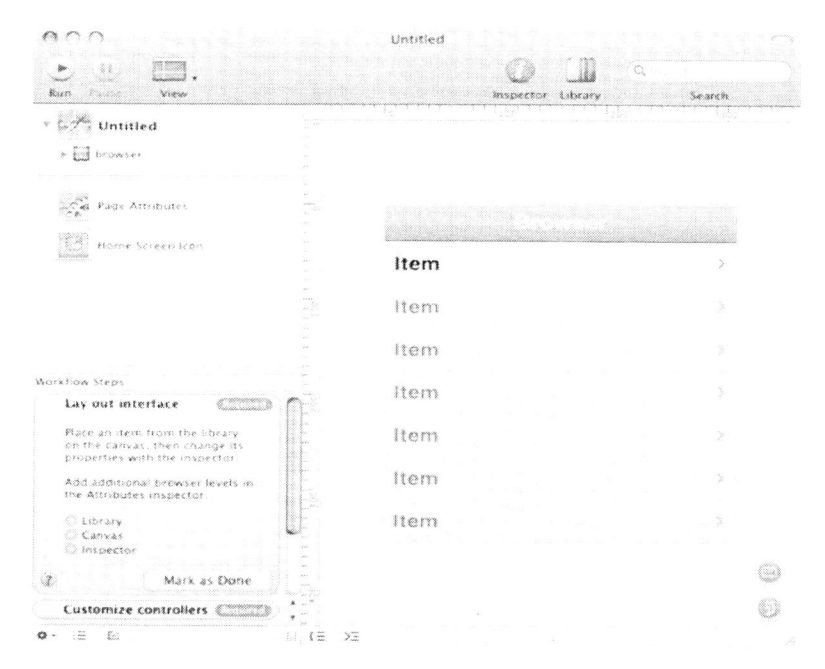

Figure 8.9. Apple™ iPhone Interface Builder.

Figure 8.10. Adobe Device Central – developing a user interface for Nokia™ Communicator 6630.

Encompassing the design decisions that have made in the first (*task-to-function-to-action design*; see Chapter 6 for further detail) and second (*action-to-effect design*; see Chapter 7 for further detail) prototypes, the final prototype design crafts a test-driven prototype along with more explicit information structure design. The key issue at this stage is to construct and envision the potential of a final prototype to make it the final product to be sustained. Hence, moving directly onto detailed prototype design and implementation with simulation is likely to commit designers to tap into specific solutions and to try out alternative design ideas. In an attempt to do so, our design process is supposed to have a series of simulation-based mock-ups of possible specific design solutions.

Simulation on emulators is technically a cheap and easy way to design mobile systems; however, emulators should be carefully used for the following reasons:

- Using screen buttons to operate an emulated device is very artificial, so the experience with the emulation will be not the same as that afforded by an actual mock-up;
- User behaviour sitting in front of a screen is very different from user behavior holding a device.

In effect, simulation on emulators can be used by developers for interactive debugging of logic, workflows, and for usability testing to understand some components of the information architecture. However, they are neither recommended for final testing with users, nor user acceptance testing.

One new alternative to this software-based emulator is to use a physical mobile platform, as reported by Nam and Lee (2003), or a video projection-based augmented reality mockup (Nam, 2005). Here, its only functionality is provided by a few buttons linked back to a desktop PC or directly rendered onto a foam mockup, as shown in Figure 8.11. However, this work is still at the preliminary stage and it remains to be seen whether there are certain benefits from this system that cannot be otherwise found using another.

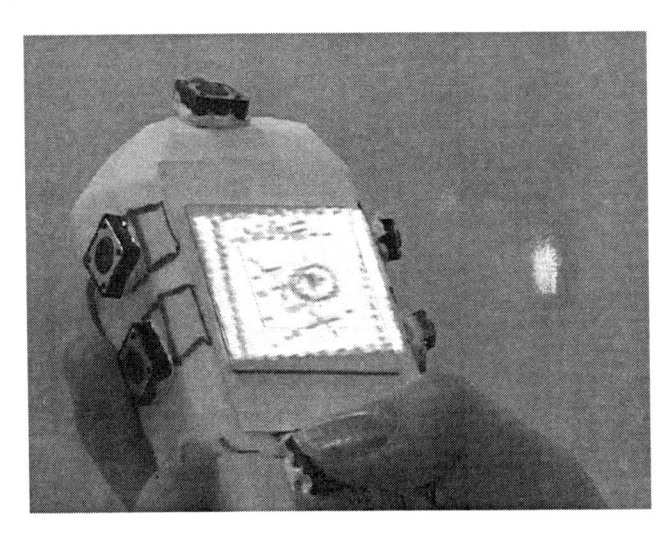

Figure 8.11. Virtual image projection onto the foam mockup. (Reprinted from Nam [2005]).

Chapter 9

COLLECTIVE WALKTHROUGHS

Successful mobile systems design requires many different considerations. Our design process proposed here dictates that we need to start by understanding the general context of the work environment. In this stage, a description of the work environment relevant to the mobile system developed is obtained for the intended user population. Based on these results, the second stage identifies a description of current major tasks or tasks to be redesigned. Several work analysis methods (i.e., task decomposition and IU scenarios) and use-case analyses have to be employed in this stage, in order to capture some representation of the knowledge that the user has, or needs to have, in order to achieve a task. In Chapters 6 through 8 we also discussed a way of developing conceptual prototypes in an iterative and piecemeal fashion considering the three aspects of interface design, i.e., *task-to-function-to-action design*, *action-to-effect design* and *information design*.

How do we mark the end of prototype design? Technically, it never ends until you release it; but common design practices will answer this question – the prototypes should be evaluated before performing full-scale usability testing. Mobile interface design is complex because it requires many different considerations that are difficult to make predictions about at an early stage of design, so inevitable is the evaluation stage that is the process of determining significance or worth, usually by the analysis and comparison of actual use progress against the assumptions by the designers, which is part of the continuing management progress of interface design.

There is no doubt that to evaluate a novel user interface, where the previous desktop-based interface evaluation is often inappropriate, one needs a new paradigm. First, the *de facto* standards that determine interface design for systems with large screens, keyboards and mice are now so familiar that the large amount of analytic and empirical HCI research that went into them should be forgotten. Second, one widely-noted feature in our design process is that designers themselves should do the evaluation, rather than recruiting users to test the conceptual design or hiring third party evaluators to do so. A key benefit of this evaluation paradigm would enable the mobile designers to justify their design decisions, and in practice to slash design lead times.

Perhaps the most common approach to evaluation is the *cognitive walkthrough* (CW: Polson et al., 1992). This is intended to be a practical and theory-based method by which a software developer could evaluate a putative user interface from the point of view of a new user of the system. The starting point for this method is a high-level task goal and a list of the actions required to complete it. The analyst is then asked to examine each action in turn by asking detailed questions with regard to sub-goal generation, recognizing the correct action from the actions possible at that point, and recognizing when a sub-goal has been completed.

Yet, as a significant drawback of CW, Blackmon et al. (2003), Cuomo et al. (1992), Hertzum et al. (2001), and Jacobsen et al. (2000) pointed out that the CW technique overemphasized task performance irrespective of the context in which the actions (or activities) occur, which limits ourselves to take performance measures only, by which we may miss crucial contextual considerations of our mobile user interface evaluation. It should be noted that this chapter does not intend to establish the "best" evaluation method, nor is it the intention to provide new insights of technical advances of the walkthrough methods. Rather, the intention is to discuss the strengths and weaknesses of each walkthrough method, so that collectively they can serve to provide a reasonable suite of usability inspection tools for mobile systems.

CONSISTENCY AND HEURISTIC EVALUATION

A mobile user interface should be evaluated prior to its release. In so doing, a practical (meaning cost-effective, but not necessarily easy-to-apply) approach – *heuristic evaluation* (HE) developed by Molich and Nielsen (1990) – can assess key areas of usability for a mobile system with the least amount of effort[62]. Although there is little consensus on what heuristics would be applicable for mobile interface design, HE enables the designer to readily review the interface and look for properties that they know, from experience, will lead to usability problems. These heuristics can offer excellent opportunities for observing how well the situated interface supports the users' work environment, if and only if the designer has had sufficient experience in heuristic-based evaluation.

A *heuristic* is a principle that is used in making a design decision. The idea behind HE is that several evaluators (perhaps designers themselves by the criteria of our putative interface design process) independently carry out a usability evaluation of a system to identify any potential usability problems with the design. One thing to notice about heuristics is that they are related to design principles and guidelines, as they make sense in evaluating a system on these principles. According to Nielsen (1994), five evaluators normally identify about 75% of the usability problems. He claimed that it is a very cost-effective process. For instance, most evaluations can be carried out in a couple of hours and each evaluator only needs a problem sheet to complete, and a copy of the system.

Practically speaking, HE depends on the evaluator's expertise. The first heuristic—*visibility of system status*—refers to the way that the system responds to each user action or to changes that the user will need to know about. This could be a message telling the user that he or she has selected a menu item; and also covers, for instance, an object moving when it is dragged, or appearing in a new screen. In a modern mobile user interface, in particular touch-based user interfaces, a lot of feedback should be designed in this way, but ensuring that it conveys the appropriate meaning takes very careful design. This system feedback must be timely, visible, and meaningful. It is crucial that the system response should be "quick," as users will soon become concerned and distracted if their actions appear to have no effect. In

---

[62] Please note that applying HE is very straightforward, but there are significant efforts required to do so, such as interpreting each heuristic for interface problems.

addition, you must be careful that the system response will be visible. It is important that the system visibly changes in a way that allows the user to instantly assess the current state of interaction—say, what the user should do next or what he or she has done just before, how they can reach the goal state and so forth. The feedback from the system must give the user such meaningful information.

A Guide for Practitioners: Heuristic Evaluations
(extended from Molich and Nelsen [1990])

- Visibility of system status – *the system should always keep users informed about what is going on, through appropriate feedback within a reasonable amount of time.*
- Match between system and real world – *the system should speak the user's language that is familiar to the user, rather than system-oriented terms.*
- User control and freedom – *users often choose system functionality by mistake and will need a clearly-marked "emergency exit" to leave the unwanted state without having to go through an extended dialogue.*
- Consistency and standards – *users should not have to wonder whether different words, situations, or actions mean the same thing.*
- Error prevention – *even better than good error message is a careful design that prevents a problem from occurring in the first place.*
- Recognition rather than recall – *make interface objects, actions, and options visible—the user should not have to remember information.*
- Flexibility and efficiency of use – *accelerators may often speed up the interaction for the expert user such that the system can cater to both inexperienced and experienced users.*
- Aesthetic and minimal design – *dialogs should not contain information that is irrelevant or rarely needed.*
- Help users recognize, diagnose, and recover from errors.
- Help and documentation.

The second heuristic, *matching between system and real world*, indicates that the system must look and behave in ways that are familiar to the user. This is particularly true when users are searching the screen looking for features with which to perform their tasks. So it is useful to think of users as searching for a match between what they are expecting and what interface objects are on the screen. To do so, we need to exploit work analysis to learn about how our intended system users describe and think about the actions and interface objects that are part of the task. One important problem for mobile interface designers is that they are very familiar with some terms that they often use in their design. In such cases, there is a danger that our design will have features that are not understandable to users, because we have not presented them in a form that the users understand. This is true, in particular, where there are many features in a modern mobile user interface.

Mobile interface designers consider the next heuristic, *user control and freedom*, for the features that are being most frequently used. With many mobile systems that we use, there is no single path for navigating (or locating/activating) through the system. Users tend to take different options (for instances, a heavy SMS user would like to customize a shortcut to sending text messages) either because they are exploring, or simply because they have made a mistake. In particular, in this error situation (refer to *error prevention*), they must be able to recover when they find themselves in unfamiliar parts of the system.

Consistency and standards are frequently appraised in every angle of evaluation, but it is hard to take measures, partly because there are few existing mobile standards, and mostly because they are not systematically possible. This *consistency* principle is concerned about supporting users' *memorable interactive cycles* from past experience and their attempts to learn the cycles as they use the system. When users use a new system, they are usually looking to understand how the system works. When they see one screen, or try an action, they are not only trying to achieve a goal, but also to remember how the whole system works. It is useful to think of learning how to use the system as learning rules of operation. The user is not just trying to understand the current action, but to gain knowledge that can be reused as he or she uses more of the system. The best a designer can do is to make sure that the number of rules that the user needs to learn is as small as possible. In relation to the number of rules, consistent design shores up the easy-to-learn. A good design is one that allows the user to use their early experience in using a few of its features again when trying more actions. A bad design will force users to learn new rules for new situations (even for situations that they already have experienced) when they could have let the user reuse something they have already learned. Also, consistency in the way that screens are arranged and features positioned is very important, as people tend to have good motor skills and spatial memory. This means that they learn and remember where interface objects or features are positioned, and use them to learn and use the system.

There are three forms of consistency in which you have to be particularly interested regarding your mobile user interface design: *visual, functional commitments* and *procedural* consistency. For *visual consistency*, all of the visual objects in a mobile user interface, such as icons, colors, and widgets, should be consistently designed and located. The visually inconsistent design might be a very small design flaw, and in actual fact, this is mostly fixed before release, thanks to corporate design guidelines and style lists or pattern libraries. By comparison, *functional commitments consistency* has seen little elaboration in either academic or industrial studies. A modern mobile phone, for instance, includes many features that users would perceive that they could apply as similar functional commitments. For instance, the task objects – audio and videos[63] – would imply commonly applicable actions, such as "play," "stop," "save," "fast forward," "rewind," "send," and so forth. Therefore, many users would think that these two would have had the same functional commitments. This functional commitments consistency can be assessed by exploiting all of the functional requirements for similar task objects (Dourish, 2001). Last, *procedural consistency* is normally built upon similar functional commitments. In order to consider the semantics of similar functional commitments, Payne and Green (1986) proposed a linguistic formalism and the multiple levels of consistency through their notational structure called *task-action grammar (TAG)*. It assumes that people organize task-action mappings according to the semantic similarities and differences between tasks. In the TAG, the *list of features* and their *possible values* give the semantic features and feature values that are used to describe simple tasks and feature lists in rewrite rules. The *dictionary of simple tasks* describes the simple tasks, for which the TAG is specified, in terms of semantic components. Also, the *task rules* prescribe the expansion of simple tasks into *primitives* and/or subtasks, the expansion of which is in turn prescribed by *subtask rules*. Finally, the *primitives* represent user actions.

---

[63] Both of them can be wrapped up as "media objects," so they may share common action features to be dealt with.

An example of the TAG notation is given in Figure 9.1. Rules (a) and (b) describe two simple tasks. Both tasks are executed by "specifying recipient," then "composing message," and then pressing "Send," except the picture message asks for another step at the beginning. Rule (c) shows how a more general rule schema can capture a consistent grammar for the two tasks.

(A) "Send a text message"
    Phone Operation[Op= compose-text-message]::=action[add-recipient],action[add message text], action[send]

    action[Op=add-recipient]::= Add Recipient[new-entry]|Add Recipient[existing-entry]| Add Recipient[multiple-existing-entries]
    action [Op=add-message-text]::=Start Text-editor[]
    action [Op=send]::=Message Operation[Op=send]

(B) "Send a photo message"
    Phone Operation[Op=compose-picture-message]::=action[add-photo],action[add-recipient],action[add message text],action[send]

    action[Op=add-photo]::=Add Photo[add-an-existing-photo]|Add Photo[take-a-new-photo]
    action[Op=add-an-existing-photo] ::= Select Existing Photo[]
    action[Op=take-a-new-photo]::=Activate Camera[]
    action[Op= add-recipient]::=Add Recipient[new-entry]|Add Recipient[existing-entry]|Add Recipient[multiple-existing-entries]
    action[Op=add-photo-message-text]::=Start Text-editor[]
    action[Op=send]::=Message Operation[Op=send]

(C) General rule schema for handing a message
    Phone Operation[Op]::=**action[object]**,action[add-recipient],action[add message text],action[send]
    action[Op=add-recipient]::=Add Recipient[new-entry]|Add Recipient[existing-entry]|Add Recipient[multiple-existing-entries]
    action[Op=add-message]::=Start Text-editor[]
    action [Op=send]::=Message Operation[Op=send]

Figure 9.1. TAG notation for the tasks "sending a photo message" and "sending a text message."

A Guide for Practitioners to Evaluate the Three Forms of Consistency

- Visual consistency – *all of the visual aspects in a mobile user interface, such as icons, colors, and widgets, should be consistently designed and located; it can be evaluated by corporate design guidelines, style lists and pattern libraries.*
- Functional commitments consistency – *similar functions (e.g., recording videos and recording sounds) should have similar system features; it can be evaluated by functional requirements analysis.*

- Procedural consistency – *for similar functions having the procedural consistency; it can be evaluated by TAG or other HCI modelling tools.*

The next heuristic, *error prevention*, recognizes that users are prone to certain types of error when they perform actions, and points to the need to cut down the chances of such errors by a thorough examination of design alternatives. Something such as a drop-down menu can be used to save users the job of typing the name of a country, and eliminate the chance of incorrect text entry, as a common example of preventing users' mishaps or errors. Many error messages do not take into account the user's perspective. In particular, they do not take into account the terms that the user is more than likely to be familiar with or understand. It is often possible to make a reasonable guess about the information that users will need once an error is made. For example, where forms are incorrectly filled out, the incorrect information is highlighted in red with an explanation of the problem. This makes it easy for the user to identify and rectify the problem that they have experienced.

Frequently, users' errors come from a situation in which they need to recall the information or commands to do the correct action. Recognition is crucial to good, efficient and satisfying user-system communication. It is the basis of display-based interaction and *learning by exploration* (Polson and Lewis, 1990). It contrasts with the concept of recall, where the user has to make a conscious effort to remember. Recognition is based on the visibility of information and the familiarity of interface objects.

The next heuristic, *flexibility and efficient to use*, is a double-edged sword that one designs in a way that gives novice users every chance to be able to use the system easily. At the same time, making the system easy for a novice may make it slow and disturbingly awkward for an expert. So we may also need to take opportunities to make interaction faster for experts. In one of the most remarkable turns in mobile user interface design, which has spun out to underlie the new direct manipulation design paradigm, graphical user interface design holds true for predictable operation by both the novice and expert users.

We should never undermine the importance of *aesthetic design*. Mobile interface designers are often tempted to fill the screen with eye-catching graphics or vivid colors. This often has the effect of cluttering up the screen, making it difficult to look at. The designer's only legitimate power is to direct user attention by making important items prominent and clear.

Last, *help systems* frequently do not enable the user to find helpful information about features and merely wastes users' time. Help can be evaluated by considering how likely a typical user is to find useful information that is relevant to what they are currently doing, rather than simply offering facts about the system.

Using these ten general heuristics is easier said than done. The challenge is the way that the designer interprets the exact evaluation practice. In many cases, it is done according to their personal design experience, so it is more than likely that HE works out well with only highly experienced designers. In actual practice, at least five HE experts conduct independent evaluation of the interface, compiling a list of ranked usability problems and paying attention to the issues of the frequency of the problem, the impact of the users, and the persistence of the problem. However, note that these heuristics are inherited from many studies in the last two decades on desktop-computing applications, so it is not straightforward to apply them for the evaluation of mobile interfaces, which means there is, as yet, no accepted list of heuristics for evaluating mobile interfaces. According to Barber (2001) and Ryu and Monk (2003b,

2006; 2005, in press), two other heuristics must be added to this list to accommodate mobile technology needs. These are appropriate mode of use and structure of information. This will be further discussed later.

Under proper circumstances, HE can be effective. However, several factors limit its use. As pinned down above, Andre (2001) pointed out that the expertise of the evaluators is of significant importance in obtaining reliable and accurate information. People with adequate experience to carry out HE evaluation on rapidly-evolving mobile technologies are scarce. Therefore, if you are going to use this method effectively, it is important that your evaluators should be experienced mobile interface practitioners. Also, HE is difficult to apply before an interface exists; consequently, any recommendations come at a late stage in development, often too late for substantive changes to be made. If the user interface specialists are not part of the development team, they may not be aware of technical limitations of the design or why certain design or implementation decisions (i.e., design rationales [MacLean, Young, Bellotti, and Moran, 1996]) were made. Technical and organizational gulfs can arise between the development team and the user interface specialists, which impede communication and correction of the problems discovered during the evaluation. Finally, according to Bailey (2001), for every correct usability problem identified by HE, there were on average almost one and a half false alarms recorded, and reporting only of a large number of low-priority problems.

A Guide for Practitioners to Perform Heuristic Evaluation

- Write an evaluation report listing all problem items: *the problem, an explanation, its cause, possible solutions, and suggestions for further investigations.*
- The report has two goals: *the formal documentation of the usability evaluation and guidelines for future projects.*
- The report should be easy to read, concise and concentrate on the evaluation results, *both positive and negative.*

In effect, beyond the advocacy of HE by its creators, little is known about how well they work—especially in mobile interface evaluation, what kinds of interface problems they are best-suited to detect, and whether developers who are not UI specialists can actually use them. That is, the critical limitation of HE in mobile interface evaluation is the non-standard format that designers can employ as of now.

Instead, early HCI modelling makes this possible in a more systematic way without the predefined heuristics as such being assumed by HE. For instance, *interaction unit walkthrough* (Ryu and Monk, in press) is a competence model that enables the designer to see whether a task in a proposed design would match those required by the intended user group. Furthermore, the designer does carry out adequate cognitive simulation that enables him or her to justify design decisions with certain details of the IU scenarios. Also, as an alternative form, e.g., *cognitive walkthrough* (CW: Polson et al., 1992; Polson and Lewis, 1990) has been favoured by many interface practitioners. It tries to structure the evaluation process so that interface developers, not UI specialists, can carry out the evaluation, potentially increasing the number of people who can do evaluations and avoiding some of the problems mentioned earlier. Although CW is an effective theory-based evaluation method, it evaluates a mobile user interface based on the knowledge states of the assumed targeted users. Consequently, if the same knowledge states are applied, the evaluation must be the same irrespective of the situations in which the activities take place. However, specifically, human activities with mobile devices naturally occur within a serious mobile context, and that context, to some extent, defines the nature of the activities. According to CW, the same performance that is executed in different contexts may be fundamentally different. Therefore, activities cannot be understood, and so should not be analyzed outside of the context in which they occur. To encompass this contextual issue in interface evaluation, some studies, e.g., Bertelsen (2004), proposed a novel approach based on *activity theory* (Engeström and Middleton, 1996; Kutti, 1996) – a.k.a. activity walkthrough (AW). These three forms of walkthroughs – interaction unit walkthrough, cognitive walkthrough, and activity walkthrough – will be further discussed in this chapter.

Yet, it should be noted that this chapter has made an important disclaimer: it does not intend to search out the "best" evaluation method among the three alternatives. Here, our intention is to discuss the strengths and weaknesses of several interface evaluation approaches (i.e., the IU walkthrough, cognitive walkthrough and activity walkthrough), so that they can collectively serve as a reasonable suite of usability inspection tools for mobile systems.

INTERACTION UNIT WALKTHROUGH

Evaluating Consistency and Congruence

As discussed above, consistency is a double-sided measure in mobile interface evaluation; namely, the benefit of consistency is evident for reducing cognitive complexity[64], but it is hard to put the concept into an evaluative practice. It is now worth studying the concept of consistency again, detail of which can be legitimately available to the analysts of mobile user interfaces.

In a consistent interface, one can easily identify and use common methods to achieve the same goals even when these goals occur in different task contexts—for instance, "OK" for confirmation wherever needed. It is assumed that these shared rules, once learned, should always be incorporated into the representation of a new task at little or no cost in training time. After a user has had some experience with the consistent user interface, learning a new task requires the acquisition of a small number of unique rules. These new rules may be a small fraction of the total number of rules necessary to execute the new task. Rules representing the common methods transfer and do not have to be relearned. In a consistent interface, these common methods can be a large part of the knowledge required to perform the next task.

By comparison, inconsistent interfaces require that users employ different methods in various contexts to achieve the common goals. Such inconsistencies can impact users in three different ways. First, inconsistent interfaces prevent transfer of existing user knowledge. Second, inconsistencies may accelerate forgetting the association between a goal and the several inconsistent methods thanks to retrospective or proactive interference. Third,

---

[64] Cognitive complexity theory (refer to CCT: Kieras & Polson, 1985), highlights the concept of congruence, proposing that the cognitive complexity of a task determines the difficulties in acquisition, transfer, and retention of the skills necessary to perform the task using a given application.

inconsistencies between old and new methods associated with a goal may cause interference in the process of learning new methods. As a consequence, retention of methods is made more difficult by the fact that they are often both arbitrary—having no logical relation to the associated goal—and meaningless. Thus, users have to memorize meaningless associations between goals and sequences of meaningless actions that accomplish the goal.

A Guide for Practitioners:
The three types of consistency, which we have discussed earlier, warrant consistent mobile user interfaces: visual, functional commitments and procedural consistency. This classification may help you take measures of evaluating consistency.

- The consistency argument for visual consistency goes as follows: when you see the "OK" button in the same place, the sensory memory of your finger can easily remember where to click it. Otherwise, you have to recognize a different location to identify which one is the "OK" and which one is not.
- The consistency argument for the second type of consistency is the user's conceptual model at the system level. The user is supposed to expect the same commitments for two similar task objects (e.g., photos and videos), so it makes sense that the two task objects should be available for sending, recording, playing and so forth.
- The consistency argument for the third type of consistency is about a dimension of how to perform the task. It is more than likely that the user would like to have a similar or the same task procedure if the functions are similar.

As an example of a consistent mobile interface, if the "OK" button is used to select a function, the same button must be used to select other functions whenever its use is needed. Yet, Smith et al. (1982) argued that there is no one right answer to consistency, emphasizing a dimension of consistency in which the dimensions happen to overlap easily. The three types of consistency that we have discussed above would be of great value in examining the dimensions of consistency.

Here, one needs to explore another concept to evaluate a mobile system, i.e., *congruence,* which gives the user a natural workflow feeling in harmony while they are performing a task with the system given. Consider a typical situation in which people use a mobile phone—for instance, setting a ringing tone to "flight mode." First, it is generally believed that users will look at the keypad, searching the control or widget of the interface that is semantically most similar to the current objective (or goal), i.e., setting a ringing tone to "flight mode." An action selection process that selects and acts on the relevant widgets follows. In such a common mobile phone use situation, there seem to be two interaction styles: *recognition-based* and *recall-based* interaction. In the former, it is believed that average mobile phone users are widely dependent on the labels, or information (e.g., icons) presented on the mobile phone interface. Such individuals, especially new or infrequent users of the mobile phone user interface, will have only abstract representations of their task and its goal structure with little or no knowledge of the consequences of their physical actions on the interface, so they have a strong tendency to select graphical widgets on the interface based on how well they semantically match one or more components of their current goal sets. In the latter, a recall-based interaction style is mostly used by very frequent, knowledgeable and competent users of the mobile user interface who already have sufficient knowledge of how to decompose a task into a collection of subtasks in order to make effective use of the relevant widgets of the interface. Therefore, the extensive knowledge of possible actions and consequences of those actions would allow them to easily apply their prior experience to current use situations. The difference between the two distinct interaction styles implies that the evaluation of *recognition-based* interaction patterns, particularly for novice users, needs to encompass the measurement of *congruence* between the tasks and their corresponding interface specifications; while the *recall-based* interaction style used by frequent users, or users who have general expectations of a well-standardized mobile phone user interface (e.g., Nokia™ phone), is more important on the issue of how their prior knowledge is readily transferable to other situations, i.e., *congruence* between user's previous knowledge and corresponding task specifications with the system given. This crucial difference indicates that a set of reasonable speculations about a user's background knowledge and state of mind while carrying out a task should be specified first, which would categorize what interaction style would be the most likely to be used. Hence, the evaluation of mobile interface should accommodate the knowledge states of targeted users, which becomes a central theme of the *interaction unit walkthrough* (Ryu and Monk, in press) below.

A Guide for Practitioners: Two Types of Congruence

- Congruence between the tasks to be performed with the system given and their corresponding interface specifications – this is particularly important for novice users.
- Congruence between a user's previous knowledge and corresponding task specifications with the system given – this is highly applicable to frequent users, so their prior knowledge is readily transferable to other situations.

In effect, note that *congruence* is a user model dominance, and inevitably one needs a user model to assess congruence. The *physical affordance* is a stronger user model in reliably choosing correct interface objects. *Semantic affordance* of the corresponding interface object forms another basis for people's understanding of what the interface objects are and how they would behave. Since people do use their existing knowledge when confronted with new situations, the design of the system must exploit that user model; consequently, we need to examine whether this exploitation has been thoroughly made in the interface given. This is especially important if people are to be able to intuit new uses for the features they have learned. For instance, a logical connection between the physical action and the virtual response means that the forms, direction and speeds of the movement should also be designed as congruently as possible—a physical counter-clockwise rotation, with jog dial, should generate a virtual counter-clockwise direction, in order to avoid *cognitive dissonance*[65].

Interaction Unit Walkthrough and Congruence

Let us revisit Figure 7.1. The communication between the user and the system can be thought of as a cycle of execution and evaluation, as explained by Norman (1988) and Monk (1998a). That is, the user formulates goals. The goals are executed on the system and the result of the action is evaluated to see if it achieves the original goal. These mental and

---

[65] An emotional state set up when two stimuli (i.e., the physical handling of the jog dial vs. its response) are inconsistent or when there is a conflict between expectation and behaviour.

physical activities that naturally communicate between the user and the system are central to the evaluative concept of *congruence*. To examine congruence, Monk (2000) reasonably breaks down the interaction between the user and the system into three paths. Figure 9.2 depicts three paths in an interactive cycle, each corresponding to an interpretation from one side to another: *goal-to-action* paths, *action-to-effect* paths, and *effect-to-goal* paths.

An interactive cycle begins with the reformulation of goals arising from the tasks or relevant visible parts of the current user interface (Ryu and Monk, 2005). In the case of "setting a ring tone," the ultimate goal can be reorganized by the interface given, i.e., "access to the menu" and then "find Setting menu item or Sound menu item." In roughly identifying their immediate subgoals from visible parts of the current mobile user interface, users will seek to take a semantically relevant action or a series of actions on the user interface for accomplishing their current goal set on the *goal-to-action* path. It is highly reliant on the repertoire of actions in the specifications of the current interface objects on the mobile user interface, e.g., buttons to click or icons to touch and so on. Accordingly, the chosen action then triggers system effects on the current system status on the *action-to-effect* path. When new system effects are presented, e.g., new screen or new menu item, the following *effect-to-goal* path deals with changes in what is perceived by the users, and then continues to generate new goals, or eliminate completed goals in their interaction context. This newly organized goal set initiates another cycle, until accomplish the original goals are accomplished. Using the fine-grained level of description of the user model assumed, an interface designer can briefly review whether or not his or her design decisions are *congruent* (Monk, 1998).

The overall goal: Setting a ring tone to "flight mode"

1. Find a best widget for the overall goal
4a. Search a menu item "Ringing tone" or similar, if the new screen is right
4b. Go back to the previous state, if the new screen is not available for the overall goal

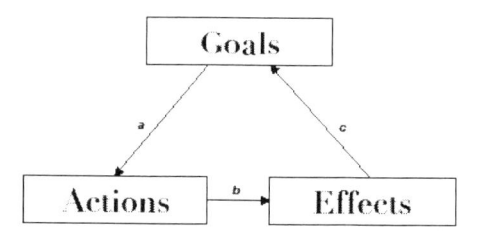

```
              Goals
              /    ^
             a      c
            /        \
           v    b     \
        Actions ---> Effects
```

2. Scan the mobile device, and then click (or touch) a most semantically similar one

3. A new screen appears (depending on system specification)

a. *Goal-to-Action problems* (Unpredictable actions), e.g., when there are no available actions for the overall goal at step 2.
b. *Action-to-Effect problems* (Unpredictable effects), e.g., when the system feedback does not include a semantically relevant menuitem in response to the action at step 3.
c. *Effect-to-Goal problems* (Unpredictable goal construction and elimination), e.g., when the system feedback does not present the next goals or acknowledge the previous goals completed at step 4.

Figure 9.2. An account of cyclic interaction for "setting a ring tone to flight mode."

The more thorough understanding that comes from the *cyclic interaction theory* makes it easier for a designer to cognitively identify interaction problems. This approach is of value because it helps the interaction designer to reason about what the cognitive process achieves, as well as what triggers that process, so the interface designer can act accordingly. Furthermore, it can help one examine design alternatives and uncover relevant design issues. Yet, it may be difficult to see the designer's own design in objective terms, and this may result in fewer problems being detected in a practical design process. This asks us to develop an analytic method – the *IU walkthrough* – to assess interaction by designers who are not HCI specialists.

The *interaction unit walkthrough* examines three types of usability problems, embracing the advantages of the interaction-perspective models, such as *CCT* (Kieras and Polson, 1985), *syndetic modelling* (Duke, Barnard, Duce, and May, 1998), and *ICO* (Palanque and Bastide, 1996). First, *action-to-effect problems* can be thought of as unpredictable effects compared with user's expectations. To set out these problems in evaluative terms, the concept of *affordance* (Djajadiningrat et al., 2002; Draper and Barton, 1993; Gibson, 1979) should be noted again. Its most basic meaning is the effects a visible (or audible) object affords. Icons, earcons, photos or widgets on a mobile user interface are often intended to allow users to perceive *semantic affordances* without forgetting once learned. Another sense of affordance is whether users can perceive how to operate interface objects (or widgets) and what to accomplish with the objects on the mobile user interface, i.e., buttons afford clicking. This *physical affordance* directly guides users' physical and logical expectations of what actions can be taken on the interface object and what effects these actions will have. In this book, affordance is considered recognition of an interface object (*physical affordance*) that can be used to take some action and the effects that will result (*semantic affordance*). For instance, the recognition of the key "6mno" on a mobile phone (in the case of the letter entry mode) allows a user to click it, which allows the user to expect a desired effect, i.e., a letter appears. Hence, any interface object used differently from its own intrinsic affordance would result in action-to-effect problems.

A Guide for Practitioners: Two Features of Affordance

- The effects that a perceptible object intrinsically affords are mostly from visible objects (e.g., icons, labels, buttons), but more recent HCI techniques also allow other sensible objects, such as audible objects (e.g., earcons, specific key-click sounds) and tactile objects (e.g., drag and drop icons onto a specific region of a mobile user interface).
- The effects that a perceptible object will generate—that is, how to operate or activate the object—and what this would generate in the system; it is in the designer's interest that this feature of affordance is the defining substance for his or her system specification.

Following on the action-to-effect path, the second type of usability problem is represented in arrow (c) of Figure 9.2. *Effect-to-goal problems* refer to whether system effects will generate any new subgoals relevant to the overall goal, or eliminate completed subgoals not required in subsequent interactions. Consider a situation in which a user would like to add her or his new contact to his or her own mobile phone. If the current user goal has a brief "and-then" goal set (Polson et al., 1992) such as "key in personal details (such as name and phone number)" and then "specify the number type" (e.g., home phone, mobile phone, relative contacts, business contacts, and so forth), but the mobile phone interface demands somewhat different expectations in the user's mental model, we see the system effects *per se* as providing subgoals inconsistent with the user's current goal set at the time of interaction. For this assessment, we need detailed assumptions of the user's current goal structure (or set) for the task being analyzed. The mobile user interface designer can use work analysis as a basis upon which to consider the adequacy of the design of task procedure, and hence the likely human performance in the system. Work analysis can thus be used successfully for assessment purposes on an existing mobile user interface as well as on one being designed.

Finally, *goal-to-action problems* (refer to Arrow (a) in Figure 9.2) in mobile user interface evaluation can be described as mapping between the current goal set assumed by work analysis, and the actions available at the time of interaction. As discussed earlier, it is generally believed that users prefer to *learn the systems by doing*, which means they will start with an abstract description of the task they want to accomplish, explore interface objects and select appropriate actions they think will accomplish the task. Hence, if the mapping between the current goal set and the actions available is not so obvious, then *goal-to-action problems* are inevitable.

On the whole, the *interaction unit walkthrough* presents a way that one can evaluate a novel user interface in terms of what psychological information is required, and what the designer can provide to create a congruent interface for the user. It is further discussed later, providing a pragmatic account of the interaction evaluation more suitable for average mobile user interface designers.

The following section proceeds by showing how IU scenarios can detect some well-known usability problems. It is organized by considering the three elements of Figure 9.2. The effect-to-goal path is what the user's mental processes must achieve in terms of perceiving an effect in order to trigger or eliminate goals. The goal-to-action path similarly specifies the link between goals and action. Finally, the action-to-effect path specifies how the system should respond. Each of these sections describes how IU scenarios could make explicit a particular interaction problem along with a general walkthrough procedure by which an analyst could evaluate an IU scenario model.

Inferring the Effects of Actions

Problems with *action-to-effect* paths are generally mode errors. The mode errors were an early concern of system modelling in HCI (e.g., Dix, 2001; Mack and Montaniz, 1994; Monk, 1986; Reisner, 1993; Tesler, 1981), and this is a further design issue in a sense that modern mobile phones have many moded interactions (e.g., camera mode, playing music mode and so forth).

An action-to-effect path exhibits moded interaction if the same action leads to different system effects under different circumstances (see example in the next paragraph). A consistent action-to-effect mapping makes learning by exploration much more effective (Ryu, 2006). It is seldom possible to remove all the modes from a mobile user interface, particularly in a mobile interface with limited input capabilities and small displays. Where modes are inevitable, the designer needs to review whether the user can easily see what the current mode is, i.e., what effect will result from some action at a particular moment. To take an everyday example, consider a remote control that can be used to work both a TV and a video player. The same buttons are used to control each appliance; which appliance responds to a particular action depends on the mode the remote is in (TV or video).

Environment			User activity		
			Mental process		Behavior
Most Recent Changes	Other Information	Current Goal	Recognition/Recall/Affordance	Change to Current Goal	Action
IU₀		Play videotape.			
IU₁ [START] TV programme.	Remote control; TV set.	Play videotape.	Affordance Push Button(VCR) -> VCR Modeln.	(+) Change mode into VCR.	Push Button(VCR).
IU₂ Nothing on both TV set and remote control; when Push Button(VCR).	TV programme; Remote control; TV set.	Change mode into VCR; Play videotape.	Recall Pushed Button(VCR); Affordance Push Button(Play) -> Videotape Play.	(−) Change mode into VCR; (+) Select Button(Play).	Push Button(Play).
IU₃ Tape Play; Tape Noise; when Push Button(Play).	TV programme; Remote control; TV set.	Select Button(Play); Play videotape.	Recognize Played Tape.	(−) Select Button(Play); (−) Play videotape.	[END]

Figure 9.3. An IU scenario including recall. If someone wants to play a video tape, she or he has to recall whether the "VCR" button was pressed last.

> *Where to look in the IU model*
> - Mode changes are detected by examining the column "Most Recent Changes" in all IUs, looking for different effects arising from the same action.
> - Where the mode does not change in the next IU, the mode information should persist in the "Other Information."
> - One should examine column "Recognition/Recall/Affordance" in the IUs, to see if the given information in "Most Recent Changes" or "Other Information" helps the user to recognize the mode signal.
>
> *How to explore action-to-effect path*
> - Step 1. Examine the "Most Recent Changes" column for instances where the same action has different effects.
> - Step 2. Examine system effects in the "Most Recent Changes" or "Other Information" columns that inform the user what the current mode is.
> - Step 3. **Hidden mode:** In Recognition/Recall/Affordance, will the user be able to recognize (rather than recall) the current mode from the system effects? If not, there is a potential hidden mode problem.
> - Step 4. **Partially hidden mode:** In the "Most Recent Change" column, are the system effects sufficiently salient for the user to discriminate the mode change arising from the previous IU? If not, there is a potential partially hidden mode problem.
> - Step 5. **Mode dependency:** Do mode signals persist in the "Other Information" column, when the current mode is the same as in the previous IU? If not, there is a potential hidden mode problem.
> - Step 6. **Misleading mode signal:** In the "Most Recent Changes" or "Other Information" columns, is it possible that mode signals imply different modes? If so, there is a potential misleading mode signal problem.

Figure 9.4. Interaction unit walkthrough – exploring action-to-effect paths with IU scenarios.

Now, assume that this mode is changed by pressing a "TV" or "VCR" button on the remote control and that no clear mode signal provided. In such an imaginary case, if you want to play a videotape you must recall that the "VCR" button was pressed last. If someone else has used the remote control since you did, this is impossible. This scenario of use is represented in IU_2 of Figure 9.3. In order to figure out the current mode, the user has to remember ("Recall") which button was pressed last. "Recall Pushed *Button(VCR)*" indicates that the designer is expecting the user to go through this mental process. *Hidden modes* are detected by examining the "Recognition/Recall/Affordance" column under the "Mental Process" columns. Where a need to recall is specified, the designer needs to be very sure that this is reasonable (because, for example, the user will always be recalling something very salient given the task and his or her previous behaviour) or think about providing some kind of mode signal.

A hidden mode is the most extreme and easy to detect version of this problem, because there are no external cues to mode. A weaker version of this is the partially hidden mode where the mode signal given has a relatively low salience. Finally, there is the possibility of misleading mode signals. For further detail on these two types of mode problems and detection of the mode problems, refer to Ryu (2003a, 2006) and Ryu and Monk (2005).

Figure 9.4 provides a general step-by-step walkthrough procedure for analyzing interaction with the IU model. Having written a complete IU model for some parts of a system, the designer is asked to examine IUs where the same action has different effects.

Goal Construction and Elimination

The effect-to-goal path is concerned with goal-construction and goal-elimination (Polson et al., 1992; Ryu, 2003b). Goal-reorganization problems can be put into four categories: (i) missing cues for goal construction – effects do not suggest appropriate intermediate goals; (ii) misleading cues for goal construction – effects suggest irrelevant intermediate goals; (iii) missing cues for goal elimination – effects do not delete completed intermediate goals; or (iv) misleading cues for goal elimination – effects delete the overall goal or intermediate goals (i.e., *post-completion error* or *super-goal kill-off* [Wharton, Bradford, Jeffries, and Franzke, 1992]). The first two possibilities result in incorrect actions or reluctance to take the correct action; the last two may result in the user unnecessarily repeating some action.

Figure 9.5 gives an example of a goal construction problem where system effects strongly imply irrelevant intermediate goals leading to incorrect actions. The phenomenon has been well explained in Bendy et al. (2000). Consider the task of copying a file onto a floppy disk. In Windows™ 2000, when users have not previously inserted a floppy disk, a dialogue box as shown in Figure 9.5a appears. Windows™ NT employs the alert box depicted in Figure 9.5b. The problem detected by the system in both cases is that the user has omitted a subgoal, that is, to insert a floppy disk. The dialogue boxes then are intended induce the user to construct this subgoal. The designer of Windows™ 2000 minimizes the actions required of the user by making this the only action required. Unfortunately, this overlooks the strong *affordance* of the sole clickable object, the "Cancel" button, and there is a tendency for users to dismiss the alert without really taking in what is required of them. By contrast, Windows NT™ (see Figure 9.5b) introduces an extra action: the user has to press "Retry" after inserting the disk. This communicates the new subtask more clearly by emphasizing its remedial nature.

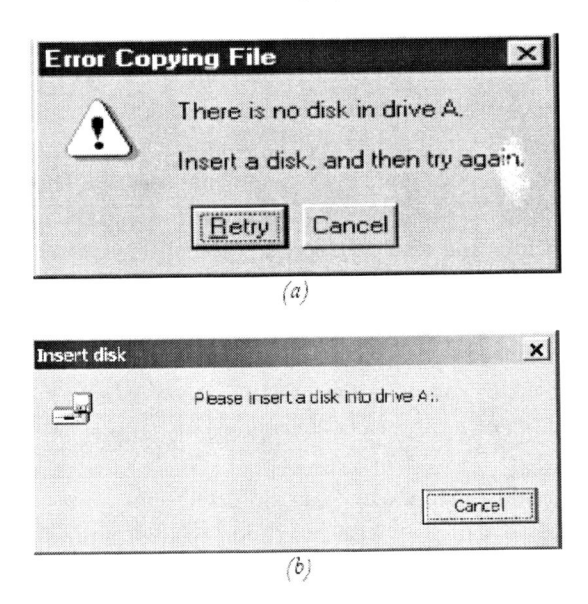

Figure 9.5. Copying a file onto a floppy disk in. (a) Windows 2000™, (b) Windows NT™.

The IUs in Figure 9.6 and Figure 9.7 illustrate how an IU analysis might have led to the detection of this problem. Looking at the column "Recognition/Recall/Affordance" in IU$_1$ of Figure 9.6, the designer assumes that the user is able to construct the subsequent goal from the message. Yet, none of system effects in the "Most Recent Change" column imply that inserting a disk will restart the copy job. Rather, the strong affordance of the "Cancel" button may make it possible for the user to click the button (note that this is partly a goal-to-action problem; see section below). In contrast, the Windows™ NT environment allows the user to plan appropriate subgoals at the cost of an extra click (see Figure 9.7). The affordance of the "Retry" button and the message informs the correct sequence of actions in IU$_1$. Figure 9.8 sets out a general procedure to check the goal-construction process in interaction design.

As well as prompting the construction of intermediate goals, system effects indicate what the system has done and thus provide feedback for goal elimination.

As with goal-construction, there are two kinds of goal-elimination problems: *missing system effects* and *misleading system effects*. A well-known example of a missing system effect is found with command-line based interface, such as disk operating systems (DOS). For instance, when a user commands "del my.doc" to delete the word file on the command-line based interface, the system responds with a new command-line prompt, if successful.

The new prompt is feedback; however, one cannot identify directly whether the file was deleted, as the feedback is provided only with respect to the lower-level goal of typing a syntactically correct command. In order for the user to eliminate the overall goal, i.e., delete a file, they have to refer back to the previous action taken – "del my.doc" – or follow up with a listing command, e.g., *dir* or *ls*, to check the right file was deleted. This goal elimination issue was further investigated in Ryu (2003b).

Misleading system effects in goal-elimination are hard to detect in the final design, in particular, in the sense that the user proactively tries to recall the correct task procedure as they make mistakes. Take Byrne's example (1995) of people leaving their cash card in an automatic teller machine after withdrawing cash. The phenomenon refers to a *post-completion error* or *super-goal kill-off* (Wharton et al., 1992). These errors are likely to occur when the final step (more specifically, an incomplete subgoal) is forgotten because a prior subgoal becomes too closely associated with the overall goal. Because they have the money, the user mistakenly believes that the overall goal has been achieved. Figure 9.9 depicts how the poor interaction design of an ATM could result in an effect-to-goal problem. Here, the overall goal (i.e., retrieve £50) has been completed at IU$_2$. As a consequence, some users may not proceed to the subsequent interactions (IU$_3$ and IU$_4$) to retrieve the card, because they have the money and therefore have accomplished the overall goal. The possibility of post-completion errors was recognized early in the design of ATMs and most now use the sequence "withdraw card, and then withdraw money" as a consequence.

In summary, goal-elimination problems can be classified into two categories: (i) implicit goal-elimination arising from missing cues; (ii) irrelevant goal-elimination arising from misleading cues. Figure 9.10 sets out a general walkthrough procedure to check the goal-elimination process in interaction design. Effect-to-goal problems arise frequently with complex procedures. "Wizards" (see Figure 7.2) reduce the problem of goal elimination by imposing a particular sequence on a task, hence providing strong guidance about what subtasks have been completed and what subtasks have yet to be completed.

Figure 9.6

	Environment		User activity			Behavior
	Most Recent Change	OI	Current Goal	Recognition/Recall/Affordance	Change to Current Goal	Action
IU_0	[START] Message(Insert Disk); Button(Cancel) Default.	Disk; Drive.	Copy a file onto floppy.	Recognize 'Please insert a disk into drive A.'.	(+) Insert disk.	Insert Disk.
IU_1		Disk; Drive.	Copy a file onto floppy.			
IU_2	DialogueBox(Insert Disk) Disappear; Disk Noise; when Insert Disk.	Disk; Drive.	Insert disk; Copy a file onto floppy.	Recognize Disk Noise.	(-) Insert disk; (-) Copy a file onto floppy.	[END]

Figure 9.6. A fragment of an IU scenario capturing the assumptions made by the designer for a copying job on Windows™ 2000.

Figure 9.7

	Environment			User activity			Behavior
	Most Recent Change	Other Information	Current Goal	Recognition/Recall/Affordance	Change to Current Goal	Action
IU_0			Copy a file onto floppy.			
IU_1	[START] Message(There is....); Button(Retry) Default; Button(Cancel).	Disk; Mouse; Drive.	Copy a file onto floppy.	Recognize 'Insert a disk, and then try again.'. Affordance Insert Disk --> Disk in Drive.	(+) Insert disk; (+) Specify 'Retry'.	Insert Disk.
IU_2	Disk Inserted; when Insert Disk.	Message(There is...); Button(Retry) Default; Button(Cancel); Disk; Mouse; Drive.	Insert disk; Specify 'Retry'; Copy a file onto floppy.	Recognize Disk Inserted; Recognize 'Insert a disk, and then try again.'; Recognize Mouse available; Affordance Click Button(Retry) --> Copy Start.	(-) Insert disk; (+) Select Retry.	Click Retry.
IU_3	DialogueBox(Error message) Disappear; Disk Noise; when Click Retry.	Disk; Mouse; Drive.	Select Retry; Specify 'Retry'; Copy a file onto floppy.	Recognize Disk Noise.	(-) Select Retry; (-) Specify 'Retry'; (-) Copy a file onto floppy.	[END]

Figure 9.7. A fragment of an IU scenario capturing the assumptions made by the designer for a copying job on Windows™ NT.

	Environment			User activity			Behavior
				Mental process			
	Most Recent Change	Other Information	Current Goal	Recognition/Recall/Affordance		Change to Current Goal	Action
IU₀			Retrieve £50.				
IU₁	[START from specifying £50 on the ATM and the card inserted]	Display(£50 highlighted); Keypad.	Specify amount; Retrieve £50.	Recognize Display(£50selected).		(-) Specify amount.	Wait Seconds.
IU₂	Message (Take out the money) Appear; Money Appear; when Wait Seconds.	Keypad.	Retrieve £50.	Recognize Message(Take out the money); Affordance Withdraw Money –> Money in hand.		(+) Take out Money.	Withdraw Money.
IU₃	Money in hand; Message(Take out your card) Appear; when Withdraw Money.	Keypad.	Take out Money; Retrieve £50.	Recognize Money; Recognize Messages (Take out the card); Affordance Withdraw Card –> Card in hand.		(-) Take out Money; (-) Retrieve £50; (+) Retrieve card.	Withdraw Card.
IU₄	Card in hand; when Withdraw Card.	Money in hand; Keypad.	Take out Card; Retrieve card.	Recognize Card.		(-) Take out Card; (-)Retrieve card.	[END]

Figure 9.9. A fragment of an IU scenario capturing the assumptions made by the designer how a user would like to use the proposed ATM. Users must recall the fact that the card has not been retrieved or looking at the machine, unless they do commit the effect-goal problem.

Where to look in the IU model
- Goal construction is detected by examining the column "Change to Current Goal" (CCG) in all IUs, looking for goals with (+).
- One should examine all of the system effects in the "Most Recent Change" or "Other Information" column to see if the system effects allow the user to construct subsequent goals that pertain to the overall goal.

How to explore goal-construction problems
- Step 1. Examine constructed goals in the CCG column in each IU for goals with (+).
- Step 2. Examine system effects in both "Most Recent Change" or "Other Information" columns in these IUs and the "Recognition/Recall/Affordance" columns.
- Step 3. **Missing cues for goal construction:** Do system effects strongly suggest the constructed intermediate goal? If not, there is a potential goal construction problem.
- Step 4. **Misleading cues for goal construction:** Do the other system effects suggest that the user conceive of goals that do not pertain to the overall goal? If so, there is a potential goal construction problem.

Figure 9.8. Interaction unit walkthrough – exploring goal construction problems with IU scenarios.

> *Where to look in the IU model*
> - Goal elimination is detected by examining the "Changes to Current Goal" column in all IUs, looking for goals with (–).
> - One should examine all the system effects in the "Most Recent Change" and "Other Information" columns to see if the system effects allow the user to perceive that the current goal has been achieved or completed.
>
> *How to explore goal-elimination problems*
> - Step 1. Examine eliminated goals in the "Changes to Current Goal" column of each IU for goals with (–).
> - Step 2. Examine system effects in both "Most Recent Change" or "Other Information" columns of these IUs that are designed to eliminate goals.
> - Step 3. **Missing cues for goal elimination:** Do system effects sufficiently allow the user to recognize (rather than recall) that the goal has been achieved? If not, there is a potential goal elimination problem.
> - Step 4. **Misleading cues for goal elimination:** Do system effects prompt the user to eliminate the overall goal even though subsequent interactions are still needed? If so, there is a potential goal elimination problem.

Figure 9.10. Interaction unit walkthrough – exploring goal elimination problems with IU scenarios.

Matching Actions to Goals

Another key aim of early usability inspection methods was to identify inadequate connections between goals and actions (Moran, 1983; Wright, Fields, and Harrison, 2000; Young, 1983; Young, Green, and Simon, 1989).

Consider Figure 9.11 and Figure 9.12. To eject a compact disk using the old version of the Macintosh desktop environment (e.g., Mac OS B1-8.6), users had to drag the disk icon to the trash can (see Figure 9.11 for an IU scenario). Novice users on the Mac environment, particularly familiar users of Windows™, may be wary of dragging their compact disk icon to the trash can icon to eject it, because this is also the action one takes to delete a file (see Figure 9.12 for an IU scenario).

Even though the two goals are rather different, the two actions are almost the same. This comes down to a problem with the affordance of the trash can. To some extent, this problem is eliminated in the new Macintosh environment (e.g., Mac OS X) by providing objects with different affordances. The icon is automatically changed into the trash can icon as a file or folder is being dragged; by contrast, it changes into the eject icon as a floppy or compact disk is being dragged onto the icon. This modification allows the user to ascertain the different affordance between dragging a disk and that of dragging a file, but not until they start to take the action. To find goal-to-action problems in a proposed design, the designer must be able to provide credible answers to the questions posed in Figure 9.13. Note that the effect-to-goal (construction) problems described above often occur partly because of the distracting nature of strong affordances of incorrect actions (see Figure 9.5a).

	Environment		User activity			Behavior
	Most Recent Change	OI	Current Goal	Recognition/Recall/Affordance	Change to Current Goal	Action
				Mental process		
IU₀			Eject a disk.			
IU₁	[START]	Icon(Floppy) Highlighted; Icon(Trash can); Mouse.	Eject a disk.	Recognize Icon(Floppy) Highlighted; Recognize Icon(Trash can); Recognize Mouse available; Affordance Drag Icon(Floppy) -> Icon(Floppy) Move.	(+) Move a selected item to trash can.	Drag Icon(Floppy).
IU₂	Icon(Floppy) and Icon(Trash can) Overlapped; when Drag Icon(Floppy).	Icon(Floppy) Highlighted; Icon(Trash can); Mouse.	Move a selected item to trash can; Eject a disk.	Recognize Icon(Floppy) and Icon(Trash can) Overlapped; Recognize Mouse available; Affordance Drop Icon(Floppy) to Icon(Trash can) -> Icon(Floppy) Disappear.	(-) Move a selected item to trash can; (+) Specify eject	Drop Icon(Floppy) to Icon(Trash can).
IU₃	Icon(Floppy) Disappear; Floppy Eject; when Drop Icon(Floppy) to Icon(Trash can).	Icon(Trash can); Mouse.	Specify eject; Eject a disk.	Recognize Floppy Ejected.	(-) Specify eject; (-) Eject a disk.	[END]

Figure 9.11. A fragment of an IU scenario capturing the assumptions made by the designer for ejecting a floppy diskette on the Mac.

	Environment		User activity			Behavior
	Most Recent Change	Other Information	Current Goal	Mental process — Recognition/Recall/Affordance	Change to Current Goal	Action
IU₀			Delete a file.			
IU₁	[START]	Icon(File) Highlighted; Icon(Trash can); Mouse.	Delete a file.	Recognize Icon(File) Highlighted; Recognize Icon(Trash can); Recognize Mouse available; Affordance Drag Icon(File) → Icon(File) Move.	[+] Move a selected item to trash can.	Drag Icon(File).
IU₂	Icon(File) and Icon(Trash can) Overlapped; when Drag Icon(File).	Icon(File) Highlighted; Icon(Trash can); Mouse.	Move a selected item to trash can; Delete a file.	Recognize Icon(File) and Icon(Trash can) Overlapped; Recognize Mouse available; Affordance Drop Icon(File) to Icon(Trash can) → Icon(File) Disappear.	[−] Move a selected item to trash can; [+] Specify delete.	Drop Icon(File) to Icon(Trash can).
IU₃	Disk Noise; Icon(File) Disappear; Icon(Trash can) Full; when Drop Icon(File) to Icon(Trash can).	Icon(Trash can); Mouse.	Specify delete; Delete a file.	Recognize File Deleted.	[−] Specify delete; [−] Delete a file.	[END]

Figure 9.12. A fragment of an IU scenario capturing the assumptions made by the designer for deleting a file on the Mac.

> *Where to look in the IU model*
> - All IUs need to be examined with regard to the affordances of their actions.
>
> *How to explore goal-to-action problems*
> - Step 1. Examine the current goal set in each IU, from the two columns "Current Goal" and "Change to Current Goal" and the IU's intended "Action."
> - Step 2. **Weak affordance of the correct action:** In the "Recognition/Recall/Affordance" column, can the user associate the action with the affordance of the corresponding object? If not, there is a potential goal-action problem.
> - Step 4. Examine the other system effects in both the "Most Recent Change" and "Other Information" columns that may have strong affordances to be taken by the user.
> - Step 5. **Strong affordance of the incorrect action:** Do system effects prompt the user to take an incorrect action? If so, there is a potential goal-action problem.

Figure 9.13. Interaction unit walkthrough – exploring interaction on the goal-to-action path with IU scenarios.

Interaction Unit Walkthrough and Other Evaluation Methods

This section looks at the overlap between Interaction Unit walkthrough and other approaches with regard to the interface problems discussed above.

There are a number of existing approaches for detecting mode problems (action-to-effect). A hidden mode is the most extreme and easy to detect version of this problem, because there are no external cues to mode and other existing approaches, e.g., *agent partitioning theory* (APT: Reisner, 1993) and *action-effect rules* (Monk, 1990), can relatively easily detect this problem.

To some extent, Polson et al.'s *cognitive walkthrough* (CW's 2.7, p. 753, 1992) can locate this type of action-to-effect problems as *time-out* moded interaction (Marila and Ronkainen, 2004). The partially hidden mode problem and misleading mode signals identified in IU scenario analysis cannot be explicitly detected by the use of APT or action-effect rules. Instead, APT provides a schematic set to describe consistency in action-to-effect paths when an action occurs in more than one rule and in the case that the rule can be abstracted.

Effect-to-goal problems have received little consideration in other techniques. CW identifies goal construction problems through a question about the "and-then" goals required. IU analysis takes a much more systematic approach; the process described in Figure 9.8 is to examine all of the effects that can prompt the construction of subsequent goals to determine whether there are missing or misleading cues.

None of the existing approaches explicitly consider the two types of goal-elimination problems discussed above. CW can detect goal elimination problems through a question about the "super-goal kill-off." IU scenario analysis, as shown in Figure 9.10, detects these problems more directly and more generally by examining subgoals that are eliminated in each IU and then determining whether all the eliminations arise from system effects.

	ACTION-TO-EFFECT			EFFECT-TO-GOAL		GOAL-TO-ACTION
	Hidden mode	*Partially hidden*	*Misleading*	*Goal construction*	*Goal elimination*	*Inappropriate affordance*
A-E rules	√	✗	✗	✗	✗	✗
CW	√ Timed mode	✗	✗	√	√	√
TAG	✗	✗	✗	✗	✗	√ Task-Action consistency
APT	√ Consistency	√ Consistency	√ Consistency	✗	✗	√ Consistency
IU	√	√	√	√	√	√

Figure 9.14. A comparison of interaction analysis methods.

The goal-to-action part of the interaction cycle was the concern of early work on interface design (e.g., Payne and Green, 1986; Reisner, 1993; Young, 1983). Payne and Green's (1986) *task-action grammar*, for example, allows the developer to ensure that similar goals are accomplished by semantically similar action sequences, thus establishing a predictable relationship between the goal (or task) and action. This is a many-to-many mapping, i.e., semantic similarity among many goals (tasks) and their corresponding actions. In a similar vein, Reisner's APT helps the designer see if a given task can be described by the same semantic features for both the designer and the user. IU scenario analysis does not include any way of explicitly capturing the semantic similarity of goal-to-action associations, but it would be possible to consider extending the procedures suggested in Figure 9.13 to include a step in which the analyst is asked to make this kind of judgment. This would involve sorting the IUs into groups according to the similarity of their goals, using the criteria suggested by Payne and Green (1986), and then examining the actions in each group for dissimilarity. While the other analyses suggested here can be carried out on fragments of an interface design, this would require a complete specification of the user interface, which may not be practical in all circumstances.

The *goal-to-action* problems considered above are easier to identify, as they are one-to-many mappings (one goal to many alternative actions). IU walkthrough asks the developer to consider the affordances of the available system effects to ensure that the correct action can easily be identified by the user and that incorrect actions will not be selected by mistake. CW (Payne and Green, 1986; Polson et al., 1992) takes a very similar in approach to IU walkthrough; here, however, we would claim that the IU notation allows for a much more systematic and explicit search for inappropriate affordances.

Figure 9.14 summarized the approaches reviewed above. As discussed above, existing approaches can analyze some but not all of the usability problems the IU approach can.

IU walkthrough also has the advantage of integrating the search for the three kinds of interaction problems around a single notation; as a consequence, it allows the designer to experience a tentative *design-evaluation cycle* at the very early design stage. That is, the designer with IU scenarios can assess their design intent from an evaluation view while he or she is designing a prototype, which can readily be a benefit in deriving the final quality prototype.

This section has described how interaction unit walkthrough provides a pragmatic approach to mobile interaction evaluation. By interaction modeling, we mean a notation that describes cognitive and environmental features at the lowest level of description. Each interaction unit specifies one step in the cycle of interaction: the visible system state that leads the user to take some action, the state of the goal stack at the start and end of the unit, and the mental processes (recall, recognition or affordance) required. This makes it possible to evaluate fragments of dialogue as they are designed by making explicit the assumptions of the designer about how the users accomplish a task.

This ability to examine fragments of human-machine dialogue design was illustrated in this section where IU scenarios are presented for some well-known problems in interaction design along with procedures for detecting various classes of problems in IU scenarios. A claim for completeness is made by considering these common problems systematically within an interaction cycle where actions lead to effects, effects to goals and goals to actions. No HCI model can cover all design issues. IU scenarios work at a high level of abstraction of both system behaviour and the cognitive processing assumed of the user and thus add what we believe to be a designer-friendly notation. This analysis thus leads to the identification of a class of usability problems that IU scenarios cannot easily detect, namely ensuring that similar goals are accomplished by semantically similar action sequences in order to establish a predictable relationship between the goal (or task) and action (Payne and Green, 1986).

Another question is how IU walkthrough will scale up to handle complex real mobile interface designs. Building IU scenarios requires effort from the designer. If the designer is developing an interface based on a well-established set of interaction methods (e.g., Windows™) there may not be sufficient benefits to justify this cost. However, in mobile user interface design, where there is a strong need to design innovative interaction techniques, the designer is faced with a device with a small display and new input methods. The hard-won, but now well-established, interaction techniques for the graphical user interfaces used on personal computers with keyboard mouse and a large screen are not easy to extend to these devices. Designers are thus returning to first principles and analysis to devise new interaction methods. IU scenarios can be used to model these more difficult design problems as fragments of interaction and there may be no great advantage to providing a complete set of scenarios for the design. We have as yet to gain experience of IU walkthrough in real design projects, and so quite how this will work out is an empirical question. However, IU scenarios are presented here as a descriptive tool that captures the minimum detail required by designers of nascent technologies and to reason about a wide range of novel user interfaces.

Perhaps the closest approach to IU scenarios is the *cognitive walkthrough* (Polson et al., 1992). This is intended to be a practical and theory-based method by which a software developer could evaluate a putative user interface from the point of view of a new user of the system. The starting point for this method is a high-level task goal and a list of the actions required to complete it. The analyst is then asked to examine each action in turn by asking detailed questions with regard to sub-goal generation, recognizing the correct action from the actions possible at that point and recognizing when a sub-goal has been completed. No notation is proposed to record the assumptions made, beyond the action list and the analyst's informal responses to their questions. IU walkthrough can thus be positioned between the formal mathematical *syndetic* modelling approach (Duke et al., 1998) and the very informal cognitive walkthrough approach. Our contention is that our relatively informal tabular notation will be easy for mobile developers to use and that they will find it useful when examining the assumptions made during mobile user interface design. IU analysis can thus complement CW by covering these weaknesses that are particularly noticeable in the case of developing new low-level interaction methods. This will be further discussed in the next section.

COGNITIVE WALKTHROUGH

A popular technique in interface evaluation is the use of the *cognitive walkthrough* method, which combines software walkthroughs with a cognitive model of *learning by exploration*[66]. One needs to note that the essence of the cognitive walkthrough approach is to take a most-likely hypothetical process, i.e., walking through procedural human information processes, and provide an evaluation at each action stage of how people actually use the system. With this hypothetical user's mental process, the designers of an interface walk through the interface step-by-step in the context of core tasks a typical user will need to accomplish. The actions and feedback of the interface are compared to the user's goals and knowledge assumed, and discrepancies between the user's expectations and the steps required by the interface are noted.

Figure 9.15. The standing state of the pizza ordering system on a PDA.

An Example: Pizza Ordering Application

One of the easiest ways to envisage the nature of the cognitive walkthrough approach is to demonstrate how it can be applied to a real mobile user interface evaluation. For this reason, a PDA-based pizza ordering application was created, as shown in Figure 9.15 (an

---

[66] It refers to the capability of users to improve their learning by regularly repeating the same type of action. The increased learning is achieved through repeated practice.

initial design). The application was designed with Microsoft Visual Studio Visio™ 2007, intended for installation on a PDA device with Windows™ CE operating system. The display size is 3.5 inches (200×160 pixels).

The representative task with this mobile application is for the customer to place a pizza order via this system. I, as a designer, note that this application is a walk-up-and-use application, because many users can use the system without training.

General Procedure of Cognitive Walkthrough

While there are many variations, generally, *cognitive walkthrough* (CW) consists of two phases: a *preparatory phase* and an *analysis phase*. In the preparatory phase, the evaluators are given the basic inputs for the walkthrough. The main analytical work follows, during which the evaluators step through each action of every task being analyzed. The details of the analysis phase[67] vary with each walkthrough technique.

Preparatory Phase
In this phase, four input conditions are considered: (i) assumptions of target users, (ii) tasks, (iii) action sequence for each task, and (iv) the interface that will be subjected to analysis.

Assumptions Regarding Target Users
This may be as simple as "people who use the PDA-based pizza ordering system," but the cognitive walkthrough asks for more clarification and details, including the users' specific background or technical knowledge that can influence the users as they attempt to deal with the particular mobile user interface. The users' knowledge of the tasks and of the mobile device should both be specified, for example:

- The user has basic knowledge of handling a PDA device or can manage it with basic guidance. They have no difficulties in using a stylus pen for input. They can also cope with common errors, which are mostly Window-based dialogue boxes.
- The main user group of this ordering system is people who own a PDA with 3G network capabilities. This is the main group of users, as the ordering system will be tailored directly to them via the Internet browser on the PDA.
- Some users may have English as a second language and they may want to use this system as a helpful aid to ordering pizzas.
- Some users may want to order their meal on the way home from work so they avoid having to make a second trip out to pick their order up later or pay extra to have it delivered.
- There also may be a group of users that would prefer to order over the Internet than via the phone, but find using a home or office desktop is not as convenient or as portable as using their mobile phone would be, or they simply do not have access to the Internet via traditional methods.

---

[67] Specifically, quite different are the predefined checklists or probing questions. However, the core part remains intact.

Although these assumptions are very hypothetical, they can justify each decision in the course of walkthrough process. In effect, there are mainly two types of users taken into consideration"

- Those who are unfamiliar with the pizza ordering procedure, and have little information about the pizza names such as "pan" base, "pepperoni" topping and so forth. They have to recognize the labels and buttons to finish the task.
- Those who are very well acquainted with the procedure and ordering mechanism. They could recall the knowledge of how to place an order in a traditional way and have proper operations on the program.

Users are offered two options when they choose pizza toppings. They could choose either a typical pizza that has fixed topping compositions, or make their own pizza with available food options.

The Tasks That the Target Users Are to Perform with the System

The walkthrough also needs detailed analyses of a suite of tasks. They should be representative tasks that most of the target users would like to accomplish. However, as it is not cost-effective to analyze all of the tasks of a mobile system, the analyses should be limited to a reasonable but representative collection of primary tasks. As discussed above, the four types of representative tasks should be of interest to average mobile system designers: *frequent, important*[68], *new* and *complicated* tasks.

The most frequent task(s) would be the determinant of the usability of the whole system. In a similar vein, the complicated (or difficult) task has a crucial effect on system usability. From a learning perspective, the performance of the frequent task, even though it is difficult, can be compromised by our learning efforts. However, if the task were less frequent, the complicated task would dictate the perceptual image of the system (i.e., *system image*). For instance, as the connection with Bluetooth™ handsets is once done, the user does not need to revisit this task until it requests repairing. Therefore, if this one-off task is too complicated, and the system-guided information is not sufficient, then the user normally ends up looking for the book-length manual. A note of new tasks is needed here. Being technically savvy themselves, designers love to enhance the capabilities of a mobile product and give users more control and more options. Too often this seems to make things harder, leaving us with mobile devices with many controls, hundreds of mysterious features and book-length manuals. The designers tend not to notice when more options make an interface less usable.

In most cases, the selection of these four types of representative tasks seems be very straightforward. First of all, the designers have great deals of requests from sales or marketing departments about what features or functions should be in the future development. Also, the designers have their own reflection or design rationales over what they have designed, which would set in the scope of the tasks revisited for the consideration.

For demonstrating the CW method, I have simply chosen some core functions of the Pizza ordering application, that is, the basic functions that the application is intended to

---

[68] Of course, in many cases, the frequent tasks may be the same as the most important tasks, but sometimes it is not the case. We can rightfully state the most important consideration in designing an operation room for a nuclear power plant (NPP); for instance, the most important task must be the "emergency stop," which would not at all be the most frequent task.

support. Task descriptions of all selected tasks must include the necessary context, such as the content of the displays that the users are most likely to use at each action step. For our example, a representative task is examined, which is "a user orders a pizza", by making up their own customized pizza.

A Complete, Written List of the Action Sequences Needed to Complete the Task

Next, there must be a description of the sequence of actions for each task. That might be a machine-level operation, such as "click the 'OK' button," or a sequence of several simple actions that a typical user would execute as a block, such as "search the pizza type." The decision as to what level of action granularity is appropriate depends on the level of expertise of the target users, which we have already assumed above ("Assumptions Regarding Target Users"). Specifying action sequences, a crucial assumption of the walkthrough approach, has provoked much criticism. For instance, even if there may be a major problem with the mobile user interface and digressions from the correct actions may occur, the evaluation merely proceeds to the next steps, as if the correct action had been always performed. The critics of this assumption say that sometimes the wrong actions would indicate true usability problems, reflecting how to avoid the usability problems in a way that the users are not expecting (Ryu and Monk, 2004a). This concern is indicative of the power of the IU walkthrough discussed above. While this criticism sounds reasonable, the assumption that the users always choose the correct action guarantees wide coverage of all of the possible usability problems in evaluation. The action procedures of the representative task in our pizza ordering example are described below.

1. Choose how many pizzas you want.
2. Choose what size pizza you want.
3. Choose your pizza flavour.
4. Choose your pizza base.
5. Choose any side dishes.
6. Choose any drinks.

The Specifications of the Mobile User Interface

The walkthrough proposes that the preparatory phase should describe the prompts preceding every action required to accomplish the task being analyzed, as well as the reactions of the interface to each of these actions. For our example, in carrying out the representative task, the following specifications were given:

Step 0: Initial Standing State

The initial state of the application is shown in Figure 9.15. The screen structure is organized with tabs at the top as depicted in Figure 9.16. There are three tabs: "Bases," "Toppings" and "Sides."

Step 1

    Current system prompt: Figure 9.15
    Action(s): No action needed

System response(s): The system automatically turns to the next page, as shown in Figure 9.16.

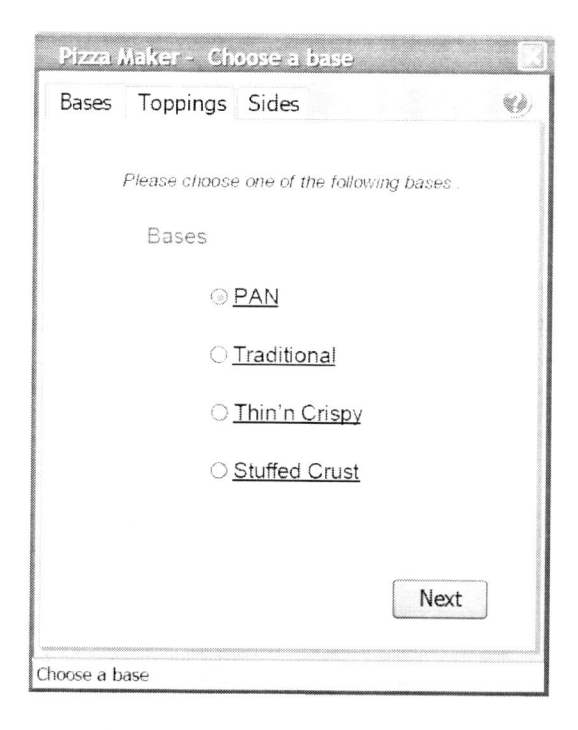

Figure 9.16. System response to the action in Step 1.

Step 2

Current system prompt: Figure 9.16

Action(s): Click one of the radio buttons to choose a pizza base[69] and then click "Next" to proceed.

System response(s): Figure 9.17

---

[69] In this case, it is assumed that the user chooses "PAN" for the base.

Figure 9.17. System response to the action in Step 2.

Step 3
　　Current system prompt: Figure 9.17.
　　Action(s): Click "Toppings" tab.
　　System response(s): Figure 9.18

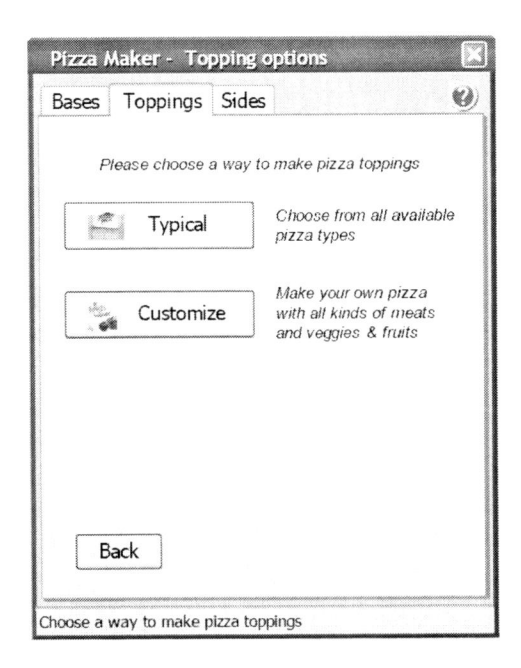

Figure 9.18. System response to the action in Step 3.

Collective Walkthroughs 199

Step 4

    Current system prompt: Figure 9.18
    Action(s): Click "Customize" button
    System response(s): Figure 9.19

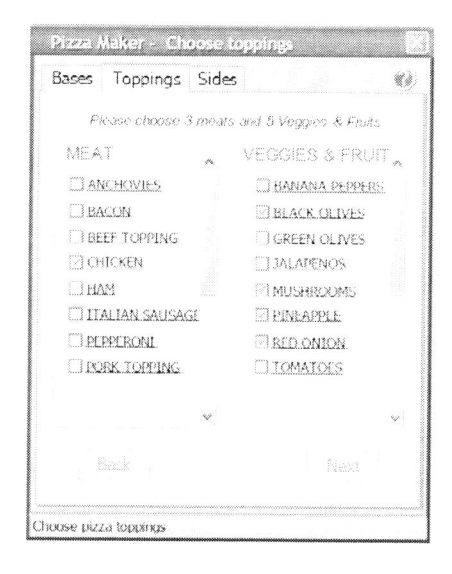

Figure 9.19. System response to the action in Step 4.

Step 5

    Current system prompt: Figure 9.19
    Action(s): Click checkboxes to select toppings as guided, i.e., "chicken," "black olives," "mushrooms," "pineapples," and "red onion" and then click "Next" to proceed.
    System response(s): Figure 9.20.

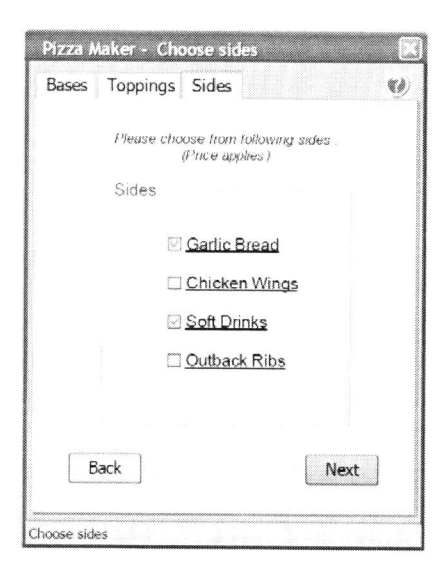

Figure 9.20. System response to the action in Step 5.

Step 6

    Current system prompt: Figure 9.20

    Action(s): Click checkboxes to select side dishes, i.e., "garlic bread," "soft drinks," and then click "Next" to proceed.

    System response(s): Figure 9.21

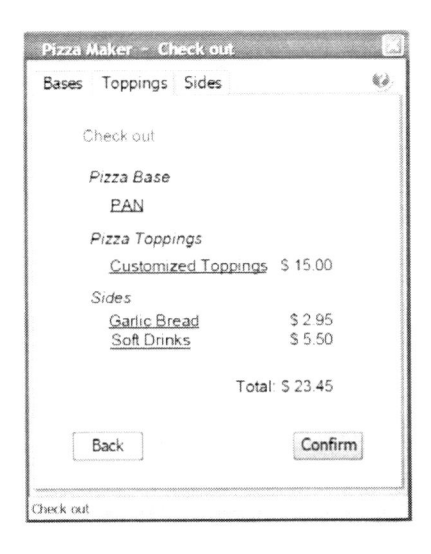

Figure 9.21. System response to the action in Step 6.

Step 7

    Current system prompt: Figure 9.21.

    Action(s): Check the details and then click "Confirm."

    System response(s): Figure 9.22

Figure 9.22. System response to the action in Step 7.

Analysis Phase

In completion of the preparatory phase, the analysis phase of the walkthrough examines each action, attempting to evaluate each action against predefined checklists or probing questions. The assessment should be based on the assumptions about the user's background knowledge and goals, which were speculated in the preparatory phase.

The predefined checklists or probing questions limit the analytic outcomes to possible usability problems with a distinct theme of usability problems. For instance, the original cognitive walkthrough (Polson et al., 1992) mainly checks if the simulated user's goals for the following actions can be reasonably assumed to lead to the next correct action, along with the three probing questions given below:

- Will the users try to achieve the right effect?
- Will the users notice the correct action or control available?
- Will the users associate the correct action with the effect trying to be achieved?

Since CW by Polson et al. (ibid.) has attempted an informal and subjective walkthrough evaluation of user interfaces, there have been many variations of the method, e.g., *cognitive walkthrough for Web* (CWW: Blackmon et al., 2003), in order to deal with some known limitations of the original CW. Yet, the core nature of CW remains intact, attempting to provide a detailed step-by-step evaluation of the user's interaction with an interface in the process of carrying out a specific task. Both the narrow focus on a single aspect of usability and the fact that the method provides a more detailed evaluation of ease-of-learning are the nature of the method's strengths and weaknesses against other model-based interface evaluations.

The evaluation comes through each action sequence, in answer to the probing questions that address the exploratory learning behaviour of the target user group. The first question—(i) "Will the users try to achieve the right effect?"—refers to whether the target users will consider the proposed right action as a logical action based on their current goal set at the time of interaction. It can be thought of as identifying *effect-to-goal* problems, in the sense that the judgment of this question is based on whether the current system status (or system effects) is fair and reasonable regarding their original task, or whether it clearly tells them what to do next (i.e., whether it gives an appropriate goal for initiating the next action). However, it is the author's belief that this many-folded probing question does not explicitly reflect the issues and concerns of mobile user interfaces, rather too implicit to related interpretations in evaluation. By contrast, the three probing questions given below are designed to identify the effect-to-goal problems, considering how the current system effects would reorganize the current goal set to maintain the subsequent interactions, particularly for mobile user interfaces.

CW for Mobile: Probing Questions for Detecting Effect-to-Goal Problems

(i) Will the system effects strongly suggest the subsequent goals?
(ii) Will the other system effects suggest that the user conceive of goals that do not pertain to the overall goal?
(iii) Will the system effects sufficiently allow the user to recognize (rather than recall) that the goal has been achieved?

The second question of the original CW – "Will the users notice the correct action or control available?" – is whether the control for the correct action is visible, audible or easily recognizable. That is, this indicates whether or not the users could experience success by their choice of proposed actions. It is the author's experience that the quality of the evaluation of the goal-to-action problems in mobile user interface is governed by the specification of the proposed actions, and affordance of the widgets corresponding to the actions. Hence, *CW for mobile* simply focuses on the affordances of the widgets on the mobile user interface, considering whether the widgets could provide appropriate affordance resulting in being selected, or not being selected, compared with the user's current goal set.

CW for Mobile: Probing Questions for Detecting Goal-to-Action Problems

(iv) Will the users associate the correct action with the affordance of the corresponding object (or widget) that is relevant to the current goal set?

(v) Will the system effects prompt the users to take an incorrect action from the strong affordance of the corresponding object (or widget) that is not relevant to the current goal set?

The final question – "Will the users associate the correct action with the effect they are trying to achieve?" – is to check action-effect problems, i.e., whether the controls can be understandable at the time of interaction. That is, if an interface presents a clear label that connects the proposed action to what users are trying to do and all the other actions seem to be wrong, the users will expect that the unique action will trigger the relevant system effects that they want to achieve. The following question is the re-statement for detecting action-to-effect problems in mobile user interfaces.

CW for Mobile: Probing Questions for Detecting Action-to-Effect Problems

(vi) Will the correct action(s) trigger the system effect(s) sufficient for the user to justify his or her action?

In a nutshell, while the original CW predominantly addresses goal-to-action problems which are their intended focus, i.e., whether the current user's goal set can clearly be accompanied by the proposed correct actions, *CW for mobile* rather emphasizes that the basic unit of walkthrough analysis should be on the level of human performance that is motivated and directed to human needs in the real context of interaction. Consequently, it systematically ensures a suite of usability problems to be tracked down with the six exploratory questions.

Having compiled the basic inputs from this preparatory phase, the evaluators step through the action sequence and tell an imaginary story about its usability, using the six questions specified above. The answers are based on the evaluator's understanding of both system specifications and the current user's goal stack. In the author's experience, an answer scheme – "Yes," "Probably yes," "Probably not," "No," and "Not applicable," with a credible story of their judgment, would be sufficient to indicate the severity of possible usability problems. The following analyses present a fragment of *CW for mobile* of the pizza ordering application conducted by the author.

Step 1

Current system prompt: Figure 9.15

Action(s): No action needed

System response(s): The system automatically turns to the next page as shown in Figure 9.16

Criteria of effect-to-goal problems

- Will the system effects strongly suggest the subsequent goals?

    Probably yes. The standing system state indicates what the system is for, i.e., make your own pizza, and automatically proceeds to the next state in two seconds. Perhaps the only concern at this interaction step is the time duration, whether the two seconds might be too long to make the user less interested.

- Will the other system effects suggest that the user conceive of goals that do not pertain to the overall goal?

    Yes. This system effect (i.e., Figure 9.15) does not indicate any other goals against the overall goal (i.e., order a pizza).

- Will the system effects sufficiently allow the user to recognize (rather than recall) that the goal has been achieved?

    Not applicable, because the overall goal has not been completed yet, and a sub-goal has not been specified yet

Criteria of goal-to-action problems

- Will the users associate the correct action with the affordance of the corresponding object (or widget) that is relevant to the current goal set?

    Probably yes. There are no widgets (e.g., buttons or labels) on which the user can take an action. Perhaps some users may not know what the correct action would be (i.e., just waiting for two seconds). To avoid this potential usability issue, the designer may consider amendments on the system effect, say, to indicate that the user does not need any further action to proceed.

- Will the system effects prompt the users to take an incorrect action from the strong affordance of the corresponding object (or widget) that is not relevant to the current goal set?

    No. There are no widgets (e.g., buttons or labels) that the user can choose that may lead to a wrong action.

Criteria of action-to-effect problems

- Will the correct action(s) trigger the system effect(s) sufficient for the user to justify his or her action?

    Yes. Figure 9.16 illustrates that the user has entered into the ordering process.

Step 2

Current system prompt: Figure 9.16

Action(s): Click one of the radio buttons to choose a base[70] and then click "Next" to proceed.

System response(s): Figure 9.17

---

[70] In this case, it is assumed that the user chooses "PAN" for the base.

Criteria of effect-to-goal problems
- Will the system effects strongly suggest the subsequent goals?

Probably yes. The user can easily develop the next goal (i.e., "Choose the base of the pizza") through the four radio buttons and the information "Please choose one of the following bases."

- Will the other system effects suggest that the user conceive of goals that do not pertain to the overall goal?

Probably yes. The three tabs shown at the top of Figure 9.16 allow the user to go ahead with "toppings" and "sides" options, rather than choosing the base first. Therefore, unless the task procedures are optional, the current system response (Figure 9.16) should be redesigned.

- Will the system effects sufficiently allow the user to recognize (rather than recall) that the goal has been achieved?

Yes. Figure 9.17 clearly indicates what the user has chosen in Figure 9.16.

Criteria of goal-to-action problems
- Will the users associate the correct action with the affordance of the corresponding object (or widget) that is relevant to the current goal set?

Probably yes. The radio buttons used in Figure 9.16 appear to be very evident and self-explanatory, except regarding what the label "traditional" means.

- Will the system effects prompt the users to take an incorrect action from the strong affordance of the corresponding object (or widget) that is not relevant to the current goal set?

No. The system effect shown in Figure 9.16 very clearly directs performance of the correct actions (in terms of the intended user's knowledge and background), except regarding what the label "traditional" means.

Criterion of action-to-effect problems
- Will the correct action(s) trigger the system effect(s) sufficient for the user to justify his or her action?

Yes. Figure 9.17 illustrates that the user has correctly selected the pizza base in the previous system state.

Step 3

Current system prompt: Figure 9.17
Action(s): Click "Toppings" tab.
System response(s): Figure 9.18

Criteria of effect-to-goal problems
- Will the system effects strongly suggest the subsequent goals?

Probably not. The system effects shown in Figure 9.17 may provide two competing sub-goals at this interaction step, either "Click OK" or "Click Toppings or Sides tab." It is necessary to implement a change in the system, with the result that when the user clicks the "OK" button, he or she is directly moved into "Toppings."

- Will the other system effects suggest that the user conceive of goals that do not pertain to the overall goal?

Probably not. The system effects clearly tell the user to complete two more sub-goals (such as selecting toppings and sides), which are highly relevant to the overall goal.

- Will the system effects sufficiently allow the user to recognize (rather than recall) that the goal has been achieved?

  No. Figure 9.18 does not indicate that the user has completed the previous sub-goal "select toppings tab," so there is a need for an indicator on the "Base" tab.

  Criteria of goal-to-action problems
- Will the users associate the correct action with the affordance of the corresponding object (or widget) that is relevant to the current goal set?

  Probably not. There are two competing actions ("Click OK" or "Click Toppings or Sides tab") that may give the intended user a confusing impression about what to do next.
- Will the system effects prompt the users to take an incorrect action from the strong affordance of the corresponding object (or widget) that is not relevant to the current goal set?

  Probably yes, because the two competing actions ("Click OK" or "Click Toppings tab") at this interaction step generate the same system effect as shown in Figure 9.18. Perhaps, one issue would be that the user may not know that the tab can lead to the same system effect as the OK button.

  Criterion of action-to-effect problems
- Will the correct action(s) trigger the system effect(s) sufficient for the user to justify his or her action?

  Yes. Figure 9.18 indicates that the user has entered into the toppings selection step.

Step 4

Current system prompt: Figure 9.18
Action(s): Click "Customize" button
System response(s): Figure 9.19

Criteria of effect-to-goal problems
- Will the system effects strongly suggest the subsequent goals?

  Yes. The statement (i.e., "Make your own pizza with all kinds of meats and veggies and fruits") next to the "Customize" button clearly indicates that the user can select his or her own toppings.
- Will the other system effects suggest that the user conceive of goals that do not pertain to the overall goal?

  No. This system effect (i.e., Figure 9.18) does not indicate any other goals against the overall goal (i.e., order a pizza).
- Will the system effects sufficiently allow the user to recognize (rather than recall) that the goal has been achieved?

  Yes. Figure 9.18 indicates that the user has come to the point where he or she can make his or her own pizza by the selection of toppings.

Criteria of goal-to-action problems
- Will the users associate the correct action with the affordance of the corresponding object (or widget) that is relevant to the current goal set?

  Yes. The user can easily select the "Customize" button thanks to the clear labelling next to the button, and its physical affordance (i.e., clickable).
- Will the system effects prompt the users to take an incorrect action from the strong affordance of the corresponding object (or widget) that is not relevant to the current goal set?

  No. The "Customize" button is salient enough for the user to choose the correct action.

Criterion of action-to-effect problems
- Will the correct action(s) trigger the system effect(s) sufficient for the user to justify his or her action?
  Yes. Figure 9.19 indicates that the user has entered into the toppings selection process.

Step 5

Current system prompt: Figure 9.19.

Action(s): Click checkboxes to select toppings as guided, i.e., chicken, black olives, mushrooms, pineapples and red onion, and then click "Next" to proceed.

System response(s): Figure 9.20

Criteria of effect-to-goal problems
- Will the system effects strongly suggest the subsequent goals?
  Yes. The user knows that he or she can choose three meats and five veggies or fruit toppings from the checkboxes.
- Will the other system effects suggest that the user conceive of goals that do not pertain to the overall goal?
  No. The system effect(s) does not presents other irrelevant sub-goals.
- Will the system effects sufficiently allow the user to recognize (rather than recall) that the goal has been achieved?
  Yes. The user can easily know that he or she has entered into the toppings selection step.

Criteria of goal-to-action problems
- Will the users associate the correct action with the affordance of the corresponding object (or widget) that is relevant to the current goal set?
  Yes. The correction action (choose more than one checkboxes) is not too difficult for the user who generally knows how to use this PDA-based application (this level of experience has been assumed for our target users).
- Will the system effects prompt the users to take an incorrect action from the strong affordance of the corresponding object (or widget) that is not relevant to the current goal set?
  No. There are no other widgets (e.g., buttons or labels) that the user can select that may lead to a wrong action.

Criterion of action-to-effect problems
- Will the correct action(s) trigger the system effect(s) sufficient for the user to justify his or her action?
  No. Figure 9.20 does not acknowledge what the user has chosen in Figure 9.19. There is a need for an indicator on the "Toppings" tab, if necessary.

Step 6

Current system prompt: Figure 9.20

Action(s): Click checkboxes to select side dishes, i.e., garlic bread and soft drinks, and then click "Next" to proceed

System response(s): Figure 9.21

Criteria of effect-to-goal problems
- Will the system effects strongly suggest the subsequent goals?

Yes. The system effects shown in Figure 9.20 clearly indicate that the user can select sides at different prices (the statement "Price applies").
- Will the other system effects suggest that the user conceive of goals that do not pertain to the overall goal?
No. The system effect does not present any irrelevant sub-goals.
- Will the system effects sufficiently allow the user to recognize (rather than recall) that the goal has been achieved?
Yes. The user can easily know that he or she has entered into the side selection step.

Criteria of goal-to-action problems
- Will the users associate the correct action with the affordance of the corresponding object (or widget) that is relevant to the current goal set?
Probably yes, but the price information is not available even though the statement says that additional prices apply. This would make the user reluctant to choose the sides.
- Will the system effects prompt the users to take an incorrect action from the strong affordance of the corresponding object (or widget) that is not relevant to the current goal set?
No. There are no other widgets (e.g., buttons or labels) that the user can select that may lead to a wrong action.

Criterion of action-to-effect problems
- Will the correct action(s) trigger the system effect(s) sufficient for the user to justify his or her action?
Yes. Figure 9.21 indicates that the user has made his or her own pizza or is simply finished with the ordering process.

Step 7

Current system prompt: Figure 9.21
Action(s): Check the details and then click "Confirm"
System response(s): Figure 9.22

Criteria of effect-to-goal problems
- Will the system effects strongly suggest the subsequent goals?
Yes. The user can see what he or she has ordered.
- Will the other system effects suggest that the user conceive of goals that do not pertain to the overall goal?
No. It is very evident that the user will click the "Confirm" button if the information is correct.
- Will the system effects sufficiently allow the user to recognize (rather than recall) that the goal has been achieved?
Yes. The user can easily know that he or she has completed every sub-goal to accomplish the overall goal.

Criteria of goal-to-action problems
- Will the users associate the correct action with the affordance of the corresponding object (or widget) that is relevant to the current goal set?
Yes. The "Confirm" button is very evident.

- Will the system effects prompt the users to take an incorrect action from the strong affordance of the corresponding object (or widget) that is not relevant to the current goal set?
  No.

Criterion of action-to-effect problems
- Will the correct action(s) trigger the system effect(s) sufficient for the user to justify his or her action?
  Yes. Figure 9.22 shows that the user has completed the overall goal.

Though one of the trickiest parts of the CW method is that there are no ways to evaluate different workflows that the user may take rather than the action specifications considered in the preparatory phase, empirical studies of the original CW, e.g., Cuomo and Bowen (1992), Hertzum and Jacobsen (2001), Jacobsen and John (2000), and Blackmon et al. (2003), demonstrated that CW has been especially promising. Yet, they also revealed that around 15% of usability problems, compared with usability testing with actual users, could not be detected by CW due to the wrong interpretation or judgment of each question. In actual fact, the judgments are heavily dependent on the evaluator's personal experience with the method, and are also affected by the way that the question is expressed. This is why *CW for mobile* has applied a more unfolded and thorough form of probing questions. Turning to the empirical data presented in Ryu and Monk (in press), one can see the case for the feasibility of the interaction unit walkthrough and some insights as to why it may be more effective than CW in some circumstances. The interaction unit walkthrough was devised for use by someone developing a user interface for a new device, or a device where interaction paradigms are not yet firmly established, as was the case with this mobile application. In the case of products using interaction paradigms that are well established, such as Web or desktop computer applications, the low-level questions concerning mode and goal reorganization in interaction unit walkthrough are not significant. This makes us consider the interaction unit walkthrough as a complement to CW by covering its weaknesses that are particularly noticeable in the case of developing novel user interfaces.

ACTIVITY WALKTHROUGH

CW for mobile is an effective theory-based evaluation method that is quite applicable for a practical assessment of mobile user interfaces. However, human activities with most mobile devices naturally occur in a wider context, and that context, to some extent, defines the nature of users' intrinsic activities with mobile systems. Consider again *activity theory* (AT) described in Chapter 4. Activity theorists claimed that activities cannot be understood, and so should not be analyzed and evaluated outside the context in which they occur. This claims to be somewhat true for activities with mobile systems.

To encompass this context issue in interface evaluation, several studies, e.g., Bertelsen (2004), proposed a practical approach based on AT (Engeström and Middleton, 1996; Kutti, 1996), and its evaluative terms are described as follows:

- the activities in which people are engaged;
- the nature of the tools (or artefacts) they use in those activities;
- the intentions of those activities; and
- the objects or outcomes of those activities.

While Bertelsen provided an exhaustive list of potential probing questions and issues and theoretical foundations of his activity walkthrough method, several aspects are not quite necessary for mobile system evaluation. Therefore, after considering a number of other studies, *activity walkthrough for mobile* is proposed here to serve as mobile interface evaluation as follows:

Understanding Contexts of the Activities: Current Activity Systems

Activity walkthrough for mobile begins with the understanding of relevant contexts in which current activities occur. The relevant contexts involve the current community and environment in which users' activities take place, the targeted users in the community who perform the activities, their intentions that motivate the activities, the tools (artefacts) and rules that can be used for the activities. They will help the evaluator pay attention to how users' actions are informed, how they are taken, and how users decide on the next step in the various contexts of use.

The following analyses show the descriptions of a Web-based system for university students to trade second-hand textbooks. You can see the Web site and/or try doing the walk-through yourself (www.massey.ac.nz/~hryu/ITbook). Please note in advance that our ensuing objective here is to propose a new "mobile-based" social networking service by understanding the current users' activities. Figure 9.23 illustrates the university students' current activity systems with the Web-based system.

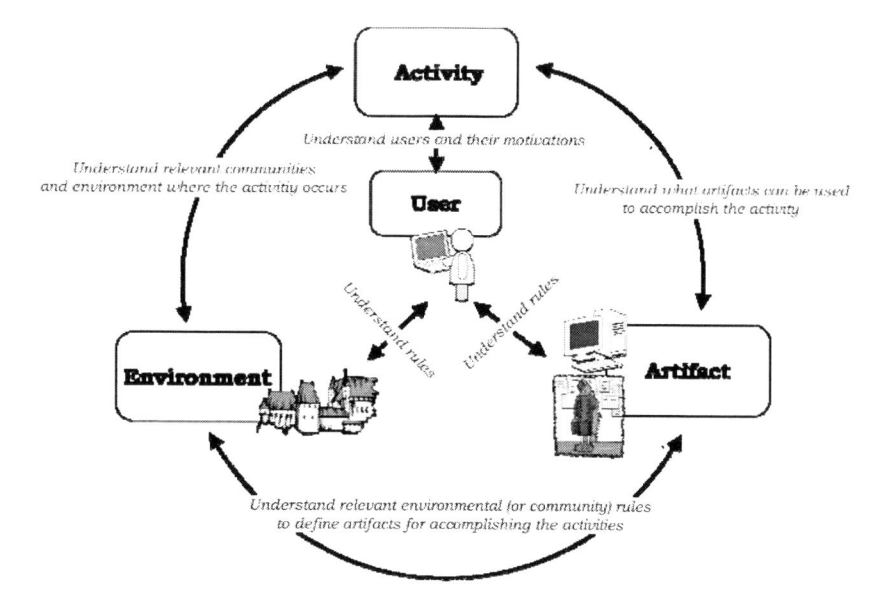

Figure 9.23. Understanding contexts of the activities: current activity systems.

Understand Relevant Environment (and/or Communities) in Which Activities Occur

A first relevant context of the activity systems is the environment or communities in which the activity systems currently work out. Targeted users for this activity system are those who have some experience with social networking services (i.e., Web 2.0 services) in New Zealand. The details are as follows:

Generally, undergraduate students at Massey University are required to buy a particular textbook for a course. All Massey students who have New Zealand citizenship or permanent residency can have student allowances from the New Zealand government, which can partially cover the cost of some textbooks. However, it is not sufficient to buy all of the new textbooks for their four to five courses in each semester. Therefore, many students are eager to buy or sell second-hand textbooks. This trading activity mostly arises prior to the beginning of every semester (January and June in New Zealand).

Currently, the university runs both a Web-based system for small advertisements and notice boards in each departmental building. The current Web page is plain-text based, so the information to be uploaded is very limited, and students must log on first with their university account. This is not a public Web site, so it is only available on campus. Therefore, most university students prefer to place their advertisements on the largest notice board at the library, with the photos of their items for sale.

Understand the Users, Their Motivations and Intentions of the Current Activity Systems

Another relevant context of the current activity systems is the users and their motivations regarding the current activity systems by which they mediate between the artefacts and environments, maintain current activities and further enhance them.

Most of the users of this activity system, in our example, are individual undergraduates. Their primary motivation is to buy or sell the second-hand books that either were used for their previous course or will be used for their future course. From the sellers' perspective, they want to sell the books that they do not need any more at the best bid and very quickly; the buyers seek to find the best quality books at the cheapest price before taking the courses.

Understand What Tools (Artefacts) Can Be Used to Accomplish the Activities

In terms of artefacts that are being used to accomplish the current activities, one is the notice board at the library and each department building. The notice board at the library is not big enough to contain all of the advertisements, so they tend to put their advertisements into the Intranet Web-based system of small ads. In addition, each department has its own small notice board in the foyer of its building, so it can be used for advertising more specialized textbooks for particular majors. When a student finds someone who wants to sell or buy, her or she must make contact to negotiate the bargain. On the other hand, the Web page under the university Web site is also being used for this trading activity. To contact someone, phone calls or emails are simply used.

Understand the Relevant Rule(s) That Can Be Used to Accomplish the Activities

Finally, equally important are the rules that mediate the relationship between the community, environment and the users. In our example, both buyer and seller are enthusiastic about buying books at a cheaper price or selling books at a higher price. To check whether the book is what they are looking for, they ask the seller the book title and the course name. If it is the one they are looking for, they begin to negotiate the price starting from the seller's bid. After mutually agreeing on a price, the buyer wants to see the book in person and the buyer and the seller meet in a particular place during the day. If the buyer is satisfied with the book status, she or he will pay the seller in cash. Normally, cheques are not acceptable. In addition, if the buyer and the seller are enrolling in the same major, they mutually check whether the seller has other books that the buyer may like to purchase.

Consolidating Contexts of the Current Activities: Proposing New Activities

Based on the understanding of the current activity systems under the relevant contexts, the next step – *contextualization* – conceptually situates new artefacts (here, it is a new mobile-based social networking service) in the context of the current activity systems by identifying whether the other entities (i.e., users, communities, environments, activities and rules) would be constantly affected by the introduction of this new artifact; more importantly, whether users' intrinsic activities in which the typical tasks are supposed to be embedded would differ from the current activity systems.

The procedure for the contextualization of our social networking service system is outlined in Figure 9.24, and detailed below.

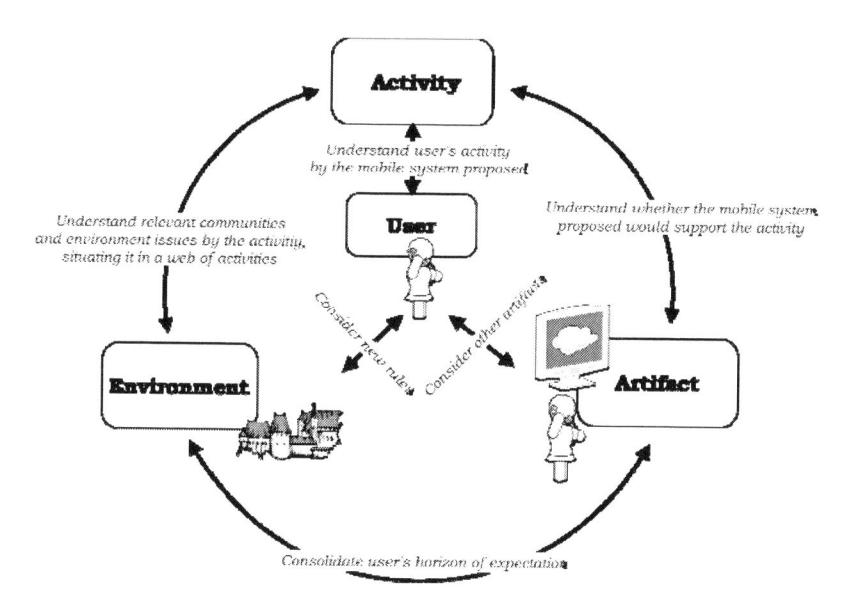

Figure 9.24. Contextualization of the new activity.

Understand the New Activities in Using the Proposed Mobile Application
The students using the social networking service system may be geared to find the books for their own courses at the university. In particular, they seek to find the cheapest and best quality book. From the seller's perspective, they want to sell their books at the highest offer and quickly. If the buyers find the book they are looking for from the new social networking

service, they contact (by the mobile application implemented on the Web site) the sellers in order to negotiate the price. When they are satisfied with the price and the status of the book, they want to buy it as quickly as they can.

Understand Whether the Proposed Mobile Application Would Support the Users' Intrinsic Activities

The potential buyers are oriented to getting the best and cheapest book when using the service. For this reason, they will search the books, contact the sellers as quickly as they can, and negotiate with them. The outcome of the sequence of actions will be to obtain the cheapest book. Conversely, the sellers want to sell their books as quickly as possible, so they will post the details (e.g., photos) of their books and decide what prices will attract possible buyers. After making contact, both parties are enthusiastic to negotiate the price in a reasonable price range. The mobile application proposed here would enhance this negotiating process, not change the intrinsic activities.

Consider Other Artefacts for Realizing the New Activities Independent of the Proposed Application

Most of the students at the university use the notice boards to trade their second-hand books around the beginning of semester. Sometimes local newspapers are used, but only if the cost of the items to be sold justifies the cost of the advertisement. Considering the wide use of Instant Messenger™ by the students, contacts with this artefact would be also possible in addition to emails or calls. However, Instant Messenger™ would not be available for someone who is off-line, so it cancels out the potential use as a new artefact.

Consider Rules of the New Activities for Analyzing Contradictions or Tensions

The social networking service mediates the seller and the buyer on the mobile-based system, thus the application should be not only usable, it should also support instant contact between the seller and the buyer. The negotiation between the seller and the buyer should be especially considered. That is, the sellers want to quickly sell their books at the best offer, and the buyers would like to buy them at the cheapest price as soon as possible.

Consolidate the User's Horizon of Expectation

To see whether the new activities systems would be acceptable by the target users, one needs to examine the user's horizon of expectation. In our example, the students have quite a lot of experience in trading and other types of social networking services, including Bebo™, Facebook™, YouTube™ and so forth. In our mobile book trading service, both parties would like to have similar functional commitments with the existing social networking services.

Verification of New Activity Systems

Interface designers can use the outcomes of the contextualization step to verify whether each task corresponds to the user's new activities in which the new artefact is going to be embedded. That is, if a particular task is not likely to take place, compared with the new activity systems, redesign of the task itself should be considered first rather than simply following the activity walkthrough analysis below.

For instance, in our social networking service, by the fact that the negotiation process via emails is not so effective in the new user's activity, the wide uptake of Instant Messenger™ (see "Understand Whether the Proposed Mobile Application Would Support the Users' Intrinsic Activities") may be considered as a new mediating artefact in carrying out the intrinsic human activities. However, Instant Messenger™ holds a significant limitation for adoption in this case, in that this artefact asks both parties to be on-line to do their negotiation process, so this task should be lifted in the new activity systems design.

Performing Activity Walkthrough for Mobile

In completion of verification of each task in terms of contextualization, the evaluators now can step through each action, using the probing questions as shown below. This walkthrough process is the same as that of the CW or interaction unit walkthrough, with the exception of the probing questions, so the demonstration is not shown here.

- Will the system effects match users' horizon of expectation so that they will be confident that progress has been made?
- Will the required machine operation make sense in the context of users' actions towards the goal?
- Will the users associate the correct machine operation with the affordance of the corresponding object?
- Will the users be able to develop matching actions in the situation?
- Will the users need instruction to be able to use the application?
- Will the machine operation match users' horizon of expectation?

Loosely speaking, *CW for mobile* leaves considerations of context and environment as afterthoughts, often considered only when a given interventions fails. Instead, *activity walkthrough for mobile* actively provides a systematic way of identifying and understanding important contextual factors in a particular situation, situating performance within the real context within which it actually takes place.

We have yet to gain experience of *activity walkthrough for mobile* in real design projects, and so quite how this will work out is an empirical question. However, the *activity walkthrough for mobile* presented here is a descriptive tool that captures the rich contextualization process and would be the most important advance against both *interaction unit walkthrough* and *cognitive walkthrough for mobile*. In particular, it provides a descriptive mechanism to easily examine various design alternatives under relevant contextual circumstances, which is not possible with the other methods.

Many mobile interface practitioners who have applied the review-based evaluation methods, such as the design principles (or heuristics) or guidelines, have suffered from substantial evaluator effects in that multiple evaluators result in a different set of problems depending on their prior experience. Also, at most, they tend to pinpoint trivial usability problems such as labels or inconsistent layout and so forth. In contrast, the task-based evaluation methods, such as *interaction unit walkthrough, cognitive walkthrough for mobile* and *activity walkthrough for mobile* allow a greater number of usability problems to be located, because the methods follow a narrow linear trajectory of analysis for each task. Hence, we have reviewed the three walkthrough methods that can be collectively used in the mobile evaluation context.

See Figure 9.25 for the summary. As noted by the case studies above, *activity walkthrough for mobile* can be of value in a very early evaluation process to obtain extensive insights of contextual usability problems; by comparison, *cognitive walkthrough for mobile* can relatively explicitly assess how the targeted user population would perform their information-seeking task with a detailed design specification. In the deepest sense of the usability problems found, *interaction unit walkthrough* is more concerned about detecting potential usability problems, gearing towards some goal by goal through a series of stages or a sequence of acts. In contrast, exploring interactions with *activity walkthrough for mobile* allows one to assess the intrinsic user's activities under use contexts, so that one can see more contextual usability issues, such as contingency, which have an impact on a user's covert or overt behaviour.

	IUW	CW	AW
Description	A method that specifies user's mental activities to assess congruent connections between goals, actions and the environment.	A method that employs task scenarios to assess the user's cognitive process	A method that include the use contexts into the evaluation to assess the intrinsic user's activities
Appropriate use	A relatively formal evaluation session for newly developing systems or functions	A relatively formal evaluation session	In a very early evaluation process to get extensive insights of the contextual reasons for possible usability problems
When to use	Design session to assess design alternatives	Relatively later design stage to check whether or not the whole task design makes sense	Very early design stage to assess appropriate functions to be implemented
Advantages	An explicit and systematic description on the intimate connection between user's action and system feedback, so that congruence on the three types of usability problems are easily detected	A deep account of how users would like to use the system, given the users' goals	A detailed explanation of the contextual issues in mobile use situation
Disadvantages	Harder to apply than CW	Lack of consideration of the context where the actions occur	May be tedious and unable to cover the entire use contexts

Figure 9.25. Appropriate use of the walkthrough methods in mobile user interface evaluation.

Although there is no hard evidence from real evaluation experience that this comparison is practically effective, several comparative studies (refer to Ryu and Monk [in press], Karat et al. [1994; 1992] and Hertzum and Jacobsen [2001]) of usability inspection methods seem to purport the accounts above.

As a concluding remark, it should be noted again that we have earlier made an important disclaimer that we do not intend to establish the "best" evaluation method, nor is it the intention to provide new insights into technical advances of the walkthrough methods. Further, it is not even possible to determine which technique is best at detecting usability problems of a mobile system, because each deals with the different scopes of usability issues. Rather, the intention is to discuss the strengths and weaknesses of each walkthrough method, so that collectively they can serve to provide a reasonable suite of usability inspection tools for mobile user interfaces.

Chapter 10

TESTING WITH USERS

Human-computer interaction concerns itself with improving the interaction between users and the computer, the mobile device in our main concern. The HCI principles and various evaluation methods that we have discussed in the previous chapter are partially addressing this issue, but a key concern still remains as to how we guarantee the quality mobile user interface by well establishing the evaluation of the usability and user experience of the interface or application. There is still a growing concern to ensure that mobile interface designers are building an interface that contains quite a few usability issues, leaving only their original design intent. The acquisition, transfer, and retention of user skills are central issues of evaluation of mobile user interfaces. New users can experience serious difficulties in learning new mobile systems. Skilled users also have problems in learning and remembering how to use new interfaces. This chapter presents pragmatic tools for testing mobile user interfaces prior to their release, as the final stage of our design process.

Academics in the field of HCI have made attempts at mapping established principles and techniques to aspects of mobile user interface testing. Most of their focus has been on identifying mobile interface heuristics. However, the most common reaction from professional mobile user interface developers has been that heuristics are too general and do not actually assist in making the mobile interface more usable. Other common approaches thus proposed by academia include using the *think aloud protocol* and *post-test surveys*.

The mobile device manufacturing industry is becoming more aware of such HCI theories and practices and is looking to integrate relevant material into their workflow and how to integrate usability testing into their development design cycle early so that it can help inform their decision-making processes. Samsung™, for instance, currently uses a combination of the *focus group study* with the *think aloud protocol* to identify key usability issues, and then focuses on these usability issues by comparing performances across participants. They also encourage all members of the development team to observe the focus group tests so that any usability issues are clearly communicated. The purpose of conducting usability testing is as such to identify and remove elements of a mobile user interface that were not intended by the designer, with the aim to end up with an interface that is exactly as the designer intended.

The general question being raised in this chapter is what the field of HCI offers mobile user interface developers to ensure better usability testing. It seems that the current tools and techniques that we employ are only useful for analyzing Web sites and software applications, but perhaps not so useful when measuring experiences with mobile devices. Indeed, some elements of a mobile user interface are very task-oriented, and perhaps traditional usability can fit in here, but it seems that we are trying to modify techniques that were designed for another purpose rather than designing techniques especially for the mobile user interface, in my personal opinion.

Usability testing is normally carried out during the design and implementation phase to find issues with the design of the application in time to fix them, answering questions like "how good is the system?," "how does the system compare to our competitors," and "what problems should we fix in the next release?" Many of the research methods and system evaluation techniques currently used in mobile user interface analysis and design are borrowed from the discipline of psychology. For example, if you are interested in finding out people's future attitude (which would not be straightforward to foretell) towards using a new mobile learning system, you would probably develop a questionnaire using questionnaire design techniques. As well as attitude measures, performance measures that are used in mobile HCI research also come from the area of experimental psychology; for example, timing how long it takes people to complete a specific task using a mobile learning system and so forth.

Yet, the main purpose of usability testing should ensure that testing is incorporated into the mobile interface design and development process, that test results are integrated into the development, and that sufficient resources are allocated to carry out the activities specified in it. What is needed, therefore, are not new techniques that are specially designed for measuring mobile user interfaces, but guidelines for mobile interface designers that help them specify what testing outcomes are meant to apply to their interface design. To assist in the process of selecting methods and techniques to be employed in the usability testing, the heuristic chart is drawn, as shown in Figure 10.1. The chart could enable one to determine which methods are appropriate for their usability testing. Here, I am trying to highlight the need to consider the demands and constraints being placed on the project that are likely to affect the appropriateness of the testing methods. The purpose of this chart is to make the decision of which method to choose based on a considered judgment, rather than an *ad hoc* selection. I am not proposing that anyone stick rigidly to the chart; rather, that designers think hard about why they have selected one method against another.

In many cases, human factors (HF) specialists are trained to do this testing job, and they should be involved, or at least consulted, in the development and execution of the usability testing. To be fair, one of the striking differences in our putative design process is that evaluation should be done by designers, but usability testing would be better done by HF specialists. This does not mean that the designers should not do testing; but the HF expertise would design and coordinate the usability tasks, ensuring that the methods by which usability criteria are assessed and put into operation are sound for the purpose of testing.

In brief, our approach makes explicit the questions that designers would ask before opting for a testing method. The factors to be considered are:

- The stage of the design cycle;
- The form of the product;
- Type of test data;
- Access to end users;
- Scales of testing.

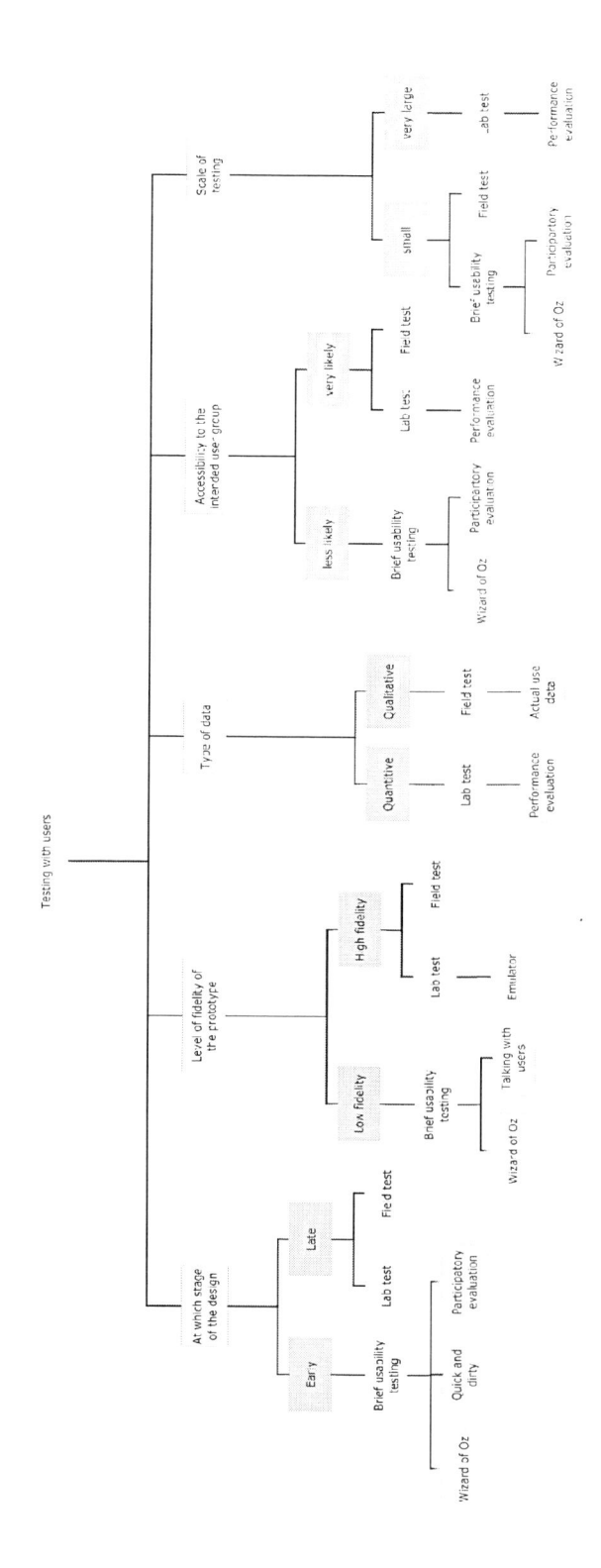

Figure 10.1. Selection of usability methods and techniques.

Also, the following are the four data acquisition methods I commonly adopt when testing an existing product:

- Talking with existing users;
- User questionnaires;
- Observation of existing users (usually in combination with talking aloud);
- Expert appraisal of the product.

Another issue to consider in usability testing is that it is not possible to test the whole range of every task to be supported by the mobile interface proposed. Care must therefore be taken in selecting tasks for testing such that they are representative of the range of tasks to be supported, and ensuring that critical task demands are met. The requirement for task representativeness can be met by selecting a suite of typical tasks, such as critical tasks (frequently most difficult and onerous), frequent (or typical) tasks, and new tasks, that cover significant aspects of the system. Frequent tasks must be included—for instance, a mobile phone must have a good task design for "calling," which is the most frequent task, so that usability of that typical task may characterize the whole usability of the system. Critical tasks are those with a certain level of risk or error or requiring too much cognitive workload to process and which therefore must meet very stringent demands on performance. Imagine a mobile in-car navigation system in which selecting and de-selecting the destination are not easy, so users must pull over their cars to change the destination. From a usability point of view, usability testing must ensure such a critical work practice throughout the test outcomes. Also, the tasks that are potentially problematic should be included in the suite of testing tasks. The list of critical, typical, erroneous, and new tasks features is compiled from the results of Stages 1 and 2 in our interface design process. Then you can list them from the highest priority to the lowest, so you can select the testing tasks based on this task priority list, too.

Finally, consider the task scenario given to users in usability testing: *abstract* or *intermediate* (such as only task structure), or *detailed task description* (including the entire sequence of actions). Depending on the different level of detail, the problems identified would vary. For instance, the abstract task scenario would ask the testers to find both the usability problems as well as the correct action sequences, so it demands more cognitive workloads of the participants. In contrast, the detailed task description can remove this cognitive workload from the testers, but they may not encounter actual usability problems from the issues associated with the task procedure because the participants in the testing may simply finish off the task without making any errors.

BRIEF USABILITY TESTING

There are basically two types of settings you can use to conduct your mobile interface testing: in the *laboratory* or in a *field*. The latter takes place in the actual environment in which people are using the mobile devices, and the former provides a controlled testing environment with a closer examination of actual usability data.

Formal usability testing (see the next section) is often time consuming and expensive. In an industrial context, many cases are called on to carry out usability testing of concepts on short notice, and deliver a result in a week or less. In such cases, neither the time nor the budget will support such a full-scale usability testing (i.e., formal usability testing). To avoid this option, expensive both in time and finance, a commonly-applied solution is to conduct an *expert appraisal*. However, it can sometimes be difficult for a mobile interface expert working in industry, simply on the basis of textbook wisdom, to argue against the potential outcomes of 15 or more participants in a formal usability testing.

The goal of *brief usability testing* is to get rapid, broad feedback on the initial design rather than exploring the whole design space. Hughes et al. (1995) proposed this type of brief usability testing, wherein you would meet with intended user groups (or sometimes non-representative users) informally and simply ask them what they think of the design. It can be formed as a very simple style of question—for instance, showing them the prototype you are currently developing and writing down any ideas or concerns they have. Brief usability testing works well with a new system sketch with which users have no previous experience. The following techniques are well perceived as pragmatic in usability testing.

Quick and Dirty Usability Testing

Quick and dirty usability testing is a compromise between the economy of expert consultation and the expense of formal usability testing. Such a test can be carried out with a few participants who, although they may not be drawn from the target population, are not technical experts and have no direct involvement with the development in progress.

My personal experience suggests that a test with five or six representative test participants will reveal major strengths and weaknesses in the usability of a product. They may complement the results of expert evaluations, which suggest that the approach has some validity. Also, the non-experts can add valuable insights and protect against the potentially limited view of experts.

The usefulness of this test is the promptness in the user interface design process, because many of the user interface designers want to get instant feedback from users during their design process.

Participatory Evaluation

One of the pragmatic methods for this quick-and-dirty usability testing is *participatory evaluation* proposed by Monk and Wright (1991). Four or six potentially intended users should be recruited to evaluate the same task together with the designer. It may be too difficult to moderate more than six users in this type of hands-on activity.

The authors suggested that there are many benefits to collecting evaluation information from several participatory users, including:

- The process is faster and effective.
- By working as a group, users must describe to each other the way they work, so extra information that users may not have thought about sharing while working alone is verbalized and unique usability issues are often identified.

I strongly recommend that you collect this participatory evaluation data from at least two groups of users. The moderator is the person who works with the users to facilitate completion of the task. This is a critical role and it can determine the success of the session, normally the designer taking this role.

Talking with Existing Users

In most cases, the existing users have experienced the best and worst aspects of the product and all we have to do is to extract their experiences and learn something from them.

The *critical incident method* (see Chapter 4) would be great in this regard. Of course, this is not as easy as it sounds, because the users are extremely adaptive and can learn to cope with the worst interfaces imaginable. In many cases, they will often have forgotten how incredibly difficult they found the product when they first encountered it. Also, in many cases, users will be proud of their mastery over the product.

The number of users available for this type of testing process is often limited. We would normally hope to talk to about a dozen users; sometimes we are forced to make do with only one. In fact, it is our experience that around four or five users usually highlight about 80% of a product's severe usability problems (Virzi, 1992).

Wizard of Oz Technique

The *Wizard of Oz* technique enables unimplemented technology to be tested by employing a user to simulate the response of a system. This technique can be used to test device concepts and techniques and suggested functionality before it is implemented.

In practice, the "wizard" sits in a hidden room, observes the user's actions, and simulates the system's responses in real time. For input device testing, for instance, the wizard will typically watch live video feeds from cameras trained on the participant's hand(s), and simulate the effects of the observed manipulations. Often users are unaware (until after the experiment) that the system was not real.

The wizard has to be able to quickly and accurately discern the user's input, which is easiest for simple for voice input or hand movements. The output must also be sufficiently simple that the wizard can simulate or create it in real time.

The Wizard of Oz technique can provide valuable information on which to indicate future designs. It can gather information about the nature of the interaction, and test which input techniques and sensing mechanisms best represent the interaction (so that subsequent efforts developing or adapting sensing technologies is appropriately directed), test the interaction of a device before building a functional model, find out the kinds of problems people will have with the devices and techniques, and investigate aspects of the product's form such as visual affordance (whether the product shows how it can be used).

User Questionnaires and Interview

After the intended user group has been exposed to the product, we can administer the *questionnaire* (see Chapter 4 for further detail) to get their perceived usability. Reflective interpretation is the major limitation.

To avoid these reflective accounts of the usability experience, often, in real testing, a one-to-one *interview* is far more effective than any questionnaire, and allows a rapport to be established with the user.

One of the practical approaches that I have used was the combined approach of user questionnaires and interviews in three different phases. The way a product is perceived at the very first encounter differs considerably from the way it is perceived after several months of daily use. In other words, the product's user interface requirements will gradually change along the so-called *continuum of use* (Snelders, Schoormans, and De Bont, 1993). Snelders et al. divided the continuum of use into three phases: the first impressions, initial use and habitual phases, so the testing should be carried out accordingly, in my personal view.

- [*Questionnaire*] The first impression phase – this phase starts when someone becomes aware of a product for the first time. The first impression of the product will create certain expectations as to what is offered (e.g., functionality, ease of use or quality image). Most of the issues during this phase deal with subjective and emotional product attributes:
    - *Labelling: does it speak for itself?*
    - *Design: does it clearly show the benefits of the product?*
    - *Overall appearance: what general impression does the product give?*

- [*Interview*] The initial use phase – this phase consists of the user's initial interactions with the product before he or she becomes familiar with using it. Typical issues during the initial phase are:
    - *Feedback: is relevant information sufficiently provided?*
    - *Self-explanation: is it immediately clear how to operate important features?*
    - *Conventions: is it relevant to what users are accustomed to?*

- [*Interview and questionnaire*] The habitual use phase – in this stage, the user's behaviour is more or less stable after he or she has used the product for some time. The emphasis during this phase is typically on functional product attributes. Characteristic issues during the habitual use phase are:
    - *Efficiency: can the product be operated with minimum effort?*
    - *Convenience: is it easy to use after a period of time?*
    - *Functionality: does it meet the user's needs?*

FORMAL USABILITY TESTING

Laboratory Usability Testing

Testing on devices in a usability laboratory provides a closer experience to the actual system. See the picture below, which captured a laboratory usability session of the "mobile ordering service" for the Grand Harbour Chinese restaurant, exemplified in Chapter 4. Yet, data collection can be a challenge – for video recording and voice recording, in particular, taking measurements are more complicated.

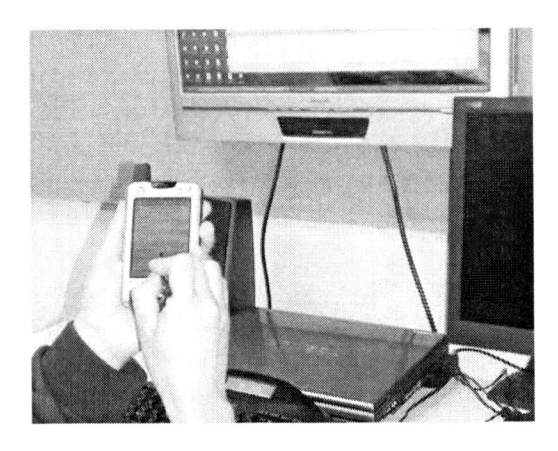

When done well, laboratory testing will capture the majority of user experience issues. It generally will not capture issues with navigation of physical and virtual spaces simultaneously, interruptions from other people or incoming interventions, or social issues, most of which will be covered by the other type of formal testing – the field test.

How Many Participants Do We Need?

This is one of the hardest questions to answer for any usability activity, and unfortunately there is no straightforward answer. This has been an area of debate within the HCI community for last two decades. If you read other textbooks about research methods, you will note that the ideal number of participants is about 30, statistically speaking. In contrast, Nielsen (1994) suggests five is sufficient; however interpretation with so few participants confuses quantitative analysis.

My personal experience reveals that at the beginning of interface development, laboratory testing with one user is enough to detect many serious usability problems, so it would be foolish to test with another user before the worst problems have been corrected. In the second testing with the interface fixed, with all of the serious problems identified from the previous testing, I normally do another testing with around two to four users. It provides a lot of observational data to keep track of. Of course, we may miss some problems that relatively few users encounter. From a practical point of view, this is the kind of risk we have to run in most projects. In the last testing session, I normally recruit around four to six users. A total of a dozen of intended users in the laboratory usability testing have been sufficient to cover around 80% of usability problems.

A Guide for Practitioners: What Kind of Users Are Needed for Testing?

The best would be the ones from the intended target user group. Otherwise, we should educate them, so it is important to find out what kind of test users we need. Their knowledge should correspond to those users we expect in practice. The user profile analysis in Stage 1 (Chapter 4) would be of great use in such cases.

The key to the laboratory test is the manner in which individuals are selected to participate, and the way those participants are subsequently assigned to group within the study has a dramatic effect on the types of conclusions that can be drawn from the test results.

A Guide for Practitioners: Where Should the Testing Be Done?

The best would be at the user's site. A mobile usability toolkit has been widely used for this reason. The advantage is that the user doesn't need to travel and the environment is the right one. The disadvantage is that we have more trouble setting up the prototype and we cannot control interruptions. The usability laboratory would be the second best option. However, many studies have shown no great difference between the two.

A Guide for Practitioners: What Tasks Should Be Tested?

As discussed above, three types of tasks must be included in the test: most frequent, most difficult and risky, and new tasks. Also, the task scenario distributed to the participants would be implicative of the test results.

Emulator Usability Testing

Emulators allow the designer to view their application on the computer, reducing the number of steps necessary to do unit testing. Emulators have an important role, i.e., software testing, but it should not be considered the real testing because of the following:

- Using screen buttons to operate an emulated device is very artificial, so the experience with the emulator will be not the same as the experience with an actual mock-up.
- User behaviour sitting in front of a screen is very different from behavior when holding a device.

An emulator usability testing is run identically to a desktop application usability testing. An emulator is likely to reveal issues with labelling, amount of information on each screen language, core user interface, softkey issues and so forth. However, it will not reveal many issues with color, contrast, light reflection, interruption and so forth, which would be in addressed in the field test. The following screenshots show the emulator usability testing of the "mobile ordering service" designed for the Grand Harbour Chinese restaurant, exemplified in Chapter 4.

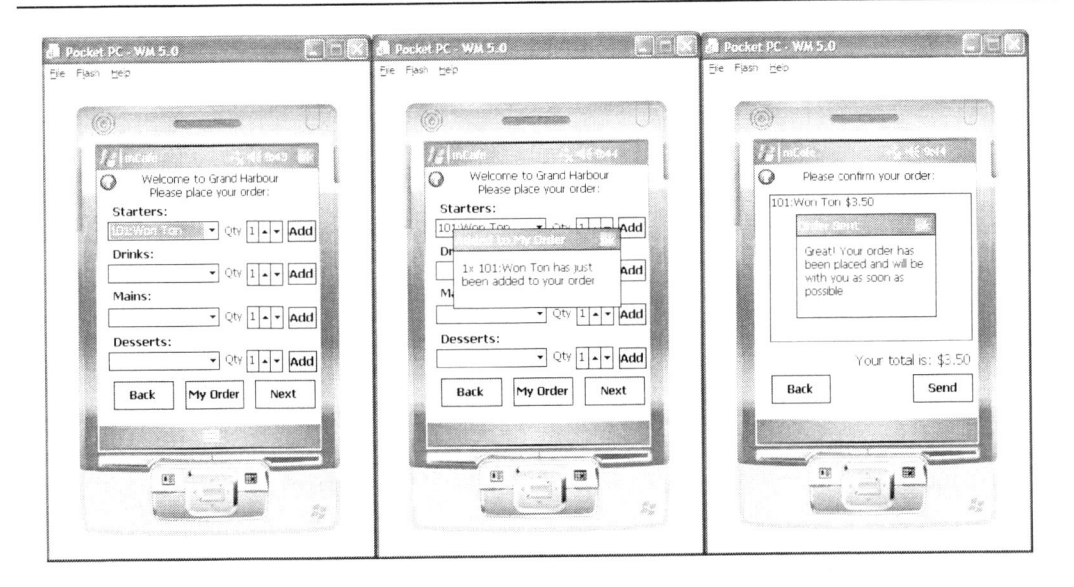

Avoiding Biases in Laboratory Usability Testing

Experts in usability testing are usually interested in predicting real usability problems that the intended user groups may encounter in their actual use cases with a small sample population, at the expense of other, perhaps irrelevant or trivial usability issues. To produce valid or meaningful and accurate conclusions, the experts must strive to eliminate or minimize the effects of extraneous influences that might detract from the accuracy of a usability testing. This section further discusses the main types of biases, and so we will have reasonably reliable usability data to ensure that the conclusions drawn are valid.

Ironically, the usability experts (in particular, those involved in observational study in laboratory testing) are the first common source of bias. Frequently called *experimenter bias*, this source of bias refers to the potential for usability experts themselves to inadvertently influence the behaviour of participants in a certain direction. That is, a usability expert who holds certain beliefs about the nature of his or her usability session and how the usability data will or should turn out may intentionally or unintentionally influence the outcomes of the study in a way that favours his or her expected outcome. These biases are particularly prevalent in usability testing in which a single usability expert is responsible for the usability testing session. A subtle example could simply be the fact that the usability expert unconsciously treats some participants differently from others, so participants might feel somewhat different after interacting with the expert prior to the usability session, and this might have an impact on their self-reporting based data (such as verbal protocols or interviews) or their attitudes toward engaging in the usability study. Another example of experimenter bias is related to the levels of knowledge in running the usability testing session. For instance, a usability expert with 20 years of experience would have more practical experience rather than someone who is just out of graduate school. Although subtle, experimenter biases can have significant impact on the validity of the usability problem found.

In both an intuitive and a practical sense, the participants can also be a significant source of bias (i.e., *participant bias*). Like usability experts, the participants bring their own unique mindset of biases and perceptions into the usability session; for instance, their attitude – is the participant anxious about the usability session, eager to please the usability expert, simply motivated by the fact that he or she is being compensated for participation, or has a preference for any particular devices? Most common references to this participant bias are the *acquiescence effect*[71] and *Hawthorne effect*[72]. Another type of participant bias is associated with the sample population. Consider that most participants in usability testing are volunteers. Accordingly, the sampled volunteer participants might be different from the general population of intended user groups as a whole, and the conclusions drawn from the usability testing might be limited to this specific sampled population. Kazdin (2003) rightfully named the different types of participants as follows:

> "The good participants might attempt to provide information and responses that might be helpful to the study, while the negativistic participants might try to provide information that might confound or undermine it. The faithful participant might try to act without bias, while the apprehensive participant might try to distort his or her responses in a way that portrays him or her in an overly positive or favourable light."

As with experimenter and participant biases, the usability expert should consider and attempt to control his or her impact on the usability testing. Conveniently, one of the methods for controlling these effects is a good experimental design through randomization (random selection and assignment). However, the discussion of this topic would be far beyond the scope of this book, so it is left to the reader to find other references (such as experimental design and statistics sources). Instead, Figure 10.2 briefly summarizes my statistical tips which offer some value for usability experts in building their own experimental design with appropriate statistical analysis.

EXPERIMENTAL DESIGN IN USABILITY TESTING	DATA TYPES COLLECTED	APPROPRIATE STATISTICAL ANALYSIS
Repeated measure design	Sign	Sign-test
	Rank	Wilcoxon signed-rank
		Friedman test
	Numerical	T-test
		ANOVA
Independent subject design	Rank	Mann-Whitney U test
		Kruskal-Wallis test
	Numerical	T-test
		ANOVA
Frequency or observation	Frequency	Fisher test or Chi-square

Figure 10.2. Choosing appropriate statistical analysis and its corresponding experimental design.

---

[71] A tendency to agree with the viewpoint of others, often with an authority. If the source is an authority, the acquiescent person will tend toward agreement regardless of the nature of the content of the statement.

[72] Participants often change their behaviour or attitudes merely as a response to being observed and to be helpful to the researcher.

Field Test

Field usability testing refers to testing outside of the laboratory. The field test, in the form of asking the user to perform both real-world activities and usability tasks simultaneously, can thus involve the greatest possibility for realistic usability problems due to the richness of the environment in which the usability tasks take place and myriad distractions that may produce effects on the performance of the usability tasks. A field test obviously does not give you control over all of your variables and settings in the same way as laboratory testing. However, it is widely believed to be an effective form of testing for mobile devices and applications under myriad dynamic social environments.

In a sense, the field test takes more time and money, but owing to its informality and the mobility of the device, the field test is believed to be more realistic than laboratory testing, and less expensive. Rather than asking potential participants to come into a laboratory to use a mobile system, take the mobile system to the participant, wherever that person may be.

Field Testing vs. Laboratory Testing

Field test settings are very different from the controlled environment used during usability laboratory testing, where tasks are set and completed in an orderly way. Instead, field tests are messy in the sense that activities often overlap and are constantly interrupted. It follows that the way people interact with products in their everyday hectic world is often different from how they perform a set of tasks in a laboratory setting. Hence, by evaluating how people think about, interact with, and integrate products within the settings where they will ultimately be used, we can get a better sense of how successful the products will be in the real world.

A common belief held by many mobile HCI researchers is that mobile interfaces need to be tested in the field, in the sense that the field test can involve realistic usability problems grounded in the relevant contexts, as opposed to the context-independent laboratory test. There are significant differences in the way to perform a laboratory testing and a field testing, in terms of both the site of the test (field vs. laboratory) and the usability tasks that the participants perform during the test. An important note for the latter is needed. Mobile system users are more than likely to be using an interface while on the move (i.e., not a static position such as in front of a computer screen on a table) or with many inherent interruptions (e.g., incoming messages while navigating the complex menu structure of the mobile phone). The extra task of navigating a physical space and interacting with surrounding objects (interruptions) and people (human-to-human interactions) presumably affects the user's ability to use the mobile interface. This implies two things in the field test. First, the usability tasks to be tested in the field are quite limited (with a smaller number of usability tasks in the test session), and should be intrinsically different from those tested in the laboratory (i.e., it is not easy to find the consistent location of the buttons in the field test). Consequently, they should be justified at the expense of cost and time. Second, since users may perform several extra real-world tasks simultaneously, the usability tasks to be tested should not require high cognitive demands or take a longer time to complete. Nonetheless, the crucial advantages are that contextual usability information over a more representative sample population of the user group can be gathered, and costs would now be very much comparable to controlled laboratory usability testing.

Despite these sound qualities of field testing, comparison of laboratory and field usability tests have shown little difference in results in terms of finding design flaws (Kaikkonen, Kekäläinen, Cankar, Kallio, and Kankainen, 2005), which may indicate that the field study that quite often consumes additional time and money should be a supplemental rather than the primary usability testing tactic. We can reasonably assume from Faulkner's (2003) findings that the variation between the field test and the laboratory test of a mobile interface was fairly marginal. Another research study done by Carlos et al. (2007) also demonstrated that conducting a time-consuming field test might not be worthwhile when searching for user interface flaws to improve user interaction.

Perhaps this can be interpreted as indicating that most of the usability problems in mobile user interface arise from the physical limitations of the interface, such as small screen size and clumsy input mechanisms; as a consequence, finding such trivial usability problems would have no such effects in both field testing and laboratory testing. Also, it is thought that users' performance in the laboratory seems to be much higher than users' performance in the field, in part because the laboratory users are aware that the experimenter observes their performance (i.e., the *Hawthorne effect*); therefore, they would perform better in finding more usability problems, mostly because they are able to focus only on the testing tasks without any of the significant interruptions that occur in everyday lives.

Considering that it is better to spend your usability testing resources on those things that can change the key performance of your mobile user interface, we need to outline four reasons for a preference for laboratory testing. Firstly, most of the usability problems found in the mobile user interface are associated with the small screen size or relatively ineffective input methods (i.e., system factors rather than use context factors), which have little to do with field testing. Also, it is not easy in the field to test specific hypotheses about an interface or account for how people react to or use a product with the same degree of certainty that we can achieve in the laboratory setting. This makes it more difficult to determine what causes users to behave in a certain way or what is problematic about the usability of a product. And most of the mobile tasks, if they are well designed, require less attention, so there would be little value in considering a lot of interruptions, by which it is assumed that field testing would result in more realistic usability problems. Finally, testers' cognitive workloads and their task performance would be a reason for concern. It can be said that the laboratory test is a very controlled way to test an interface, so it probably leads to some extent to identifying more usability problems as opposed to its counterpart.

Of course, these four reasons taken into consideration are not intended to undermine the power of the field test. The anticipated findings from the field test, which work toward resolving more than the plain usability flaws with more realistic and empirical context use, are best for the new features and functional updates. These qualitative accounts and descriptions of people's behaviour and activities from field testing reveal how they used the product and reacted to its design.

Measures

Though there are many different ways to run usability testing sessions, their common feature is that they are only interpretable when their outcomes are expressed either quantitatively or qualitatively, or both. Frequently, simple techniques such as questionnaires, observations, and interviews[73] are quite applicable in most testing sessions. This section focuses on the types of measures that you can use in testing mobile user interfaces to help you collect mobile usability data.

A Guide for Practitioners: Appropriate Measures

- More than one measure would be better, because one measure cannot reveal all of the usability problems.
- Using both quantitative and qualitative measures would be better.
- "Intention to use" or "attitude toward future use" are always a good thing to be included in the survey method (Davis, 1989).

Task Completion Time

Usability techniques can provide quantitative metrics for *time on task, task completion rate* or *how efficient the test participant is relative to an expert*, and so forth. Established work measurement methods (as a key reference, please refer to Wilson and Corlett, 1995) typically involve collecting detailed measures of time to perform elements of work.

This traditional approach to performance measurement may in some circumstances give useful approximations of usability, but they may have the clear disadvantage of relying on time data to draw inferences about work performance with as yet unimplemented systems, which will be also affected by rather different factors such as aesthetics (Kurosu and Kashimura, 1995). Many studies done by McKenzie et al. (MacKenzie and Soukoreff, 2002; MacKenzie and Zhang, 1999) have used this time measure to introduce a new mobile keypad design. However, one thing you need to be aware of is that speed of performance does not necessarily tell you how difficult the users found the task, for instance.

Errors or Revisits

The mistakes that people make when trying to complete the tasks you have assigned to them are also a good measure of performance. Errors can be recorded in a number of ways. One method would be to sit and observe people interacting with the system or service and record the number and the nature of errors as they occur. Another approach would be to automate the process if at all possible by creating a log file that records all the data input from the user in the course of a trial.

Ryu and Monk (2003b; 2005), instead of counting errors, employed a number of *revisits* to a particular place to evaluate information structure in a mobile user interface. In their articles, the participants were asked to find a certain menu item and the number of revisits to the menu item in the complex hierarchy was counted to define the usability of four different types of mobile user interfaces.

Time-Out or Silence

When mobile phones are designing the dialogues for their services, they start working on the principle that after the service delivers a prompt to a user, the user has, for instance, about three seconds[74] to make a response. If they fail to respond within three seconds, this is known as a *time out* and the system will prompt the user again for a response from scratch. This

---

[73] See Chapter 4 for further detail of these techniques.

timeout rate can therefore be defined as the number of times a user did not respond by the end of a timeout period following a system prompt divided by the overall number of responses made by a user to menu prompt from the service throughout the trial (Love, Foster, and Jack, 1997, 2000; Marila and Ronkainen, 2004).

In Ryu's (2003a) doctoral thesis, he also employed this time-out measure to see what time duration would be good enough for users to perceive, whether their goals have been accomplished or not, in several time-out cases.

Time-Series Study

Time-series study is a technique used in measuring performance over time (over a period of time). Observing and establishing the normal fluctuation of the performance over time allows the researcher to more accurately interpret the impact of system usability.

For instance, you could carry out interviews each week after participants had completed their tasks to investigate if their attitude towards the service had changed over a period of time. Taken all together, this data could provide you with useful information on the way people used the system and if some parts of the system (such as specific navigation cues) were easier to remember than others. Following the intervention, several more periodic measurements are made. There are two basic variations of this design: the simple time-series design[75] and the reversal time-series design[76].

Protocols (Verbal or Visual)

For usability testing, we can simply observe the naturally occurring behaviour of people with mobile devices. For instance, you can watch people's reactions as they interact with a new PDA interface, observing where they have difficulties or which parts of the interface they do not like, or you can also do an observation in a usability laboratory or in the field.

Indeed, observations take little effort to perform, but require a huge effort to interpret the observational data. Firstly, watching someone can have a bearing on his or her performance. Therefore, we may try to be as unobtrusive as possible, but inevitably the presence of an observer can lead people to engage in behavior beyond their normal activity—again, the *Hawthorne effect*. Alternatively, one could attempt to observe behavior covertly, i.e., by using hidden cameras. However, this also raises issues of ethics in conducting observations. Second, in observational study, there is no way for use to see inside a user's head, to see how they solve and see a problem.

A Guide for Practitioners: Observation Study

- Define clearly the aim of the observation study.
- Define the scenario in which the product will be used.
- Define the type of data to be collected.
- Define the means by which data will be presented.
- Define recording tools.

---

[74] The time duration varies among the carrier networks.
[75] A within-subjects design in which periodic measurements are made on a single group, along with no further interruption, as depicted here: Interface A (week 1) – Interface A (week 2) – Interface A (week 3).
[76] Refers to the idea that causality can be inferred if changes that occur following the presentation of an intervention diminish or "reverse" when the interface is withdrawn, as depicted here: Interface A (week 1) – Interface B (week 2) – Interface A (week 3).

The idea of using *verbal protocols* as a form of usability testing was originally put forward by Ericsson and Simon (1985). The idea behind this is that it could potentially provide you with information on what the user is actually thinking about while they are interacting with a system. That is, a user is given a problem to solve and is asked to describe (specifically, talk aloud) his or her thoughts as he or she solves it. The user interface designer can observe their mental performance by recording their comments, at times prompting or questioning. Researchers will often use some form of verbal protocol in conjunction with observational data in order to get as much information as possible from the participant in the study. For instance, by adopting this approach, you may discover what menu items are not clearly labelled (e.g., a user says aloud, "*I don't think this label is good for sending a message*") by the user talking about the problems they are having in choosing the correct menu item. In addition, it may let you know that the navigation structure is not as clear as you originally thought it was. You may also get feedback on their subjective feelings at particular points in the interaction, especially if the system goes wrong.

The verbal protocol (interchangeably, thinking aloud) asks people to talk out loud while they are completing the task you have given them. Although this is a useful form of data collection of thinking processes by monitoring a subject's verbal descriptions of what he or she is thinking at that moment, it can be a tricky process, as some people may find it difficult to talk out loud and complete a task at the same time, and other people may just feel embarrassed about the idea of this right from the start. Indeed, this would have a certain effect on task performance.

To minimize the side effect of verbal protocol in talk-aloud exercises, another form of verbal protocol data collection is the *post-event protocol*. If you made a video recording of their session, the participants could do this by going through it with you, while providing you with a running commentary of the reasons for their actions when they were carrying out the tasks. They could also provide you with information on their subjective feelings as they went through this process. If you have looked at the participants' sessions and there are specific aspects you want to focus on (e.g., a menu link that was causing most participants problems), then you could play back only these parts and ask for feedback. Monk et al. (1993) suggest that this is a very useful method, as it can prompt a participant to inform you of any particular problems that they had with the system (which they did not verbalize during the think aloud procedure) and it could also highlight particular parts of the interface that they thought was very usable.

A Guide for Practitioners: Conducting a Think-aloud Exercise

- Do a trial run – *because thinking aloud is an unfamiliar communication process to most participants, they have to get used to this rather awkward process. Before the think-aloud session, have the expert narrate his or her performance with one or two elementary tasks. The objective is to have the learner become comfortable with talking out loud.*
- When necessary, prompt the task performer to speak out – *the think-aloud process works best when the user completes the task and commentary with minimal interference from the experimenter. However, the user may forget to talk out loud, or will simply make a cursory comment. In such cases you should briefly intervene. For instance, if the user is not talking, try questions such as "what are you thinking?"*
- Review the session with the performer – *ask the performer if there were any ideas or steps that they failed to verbalize. Review your notes and ask why certain words were*

used or why they had a certain expression on their face. During the review you can ask the performer to call upon their comments or reactions, while you take notes. If you videotaped the performance, replay the video and watch it with the performer, stopping it to ask questions such as "Why did you do that?"

Verbal Protocol Analysis

Having collected and transcribed the verbal data, an analysis technique is applied, which was recommended by Gilhooly et al. (1996) and Green et al. (1996).

*A Guide for Practitioners to Use the Think-aloud Method
(adopted from Courage and Baxter [2005])*

- The key to success in using this method is practice. You must give your participants practice in using the method before data collection begins.
- One way of doing this is to get your participants to tell you how to get to this place from their home. This will help participants to get into the habit of "thinking aloud" about a task you have given them.
- The next step is to give them a practice session using the system they are going to evaluate (pilot study). If there are any silences at this point, you can gently remind them to tell you what they are thinking.
- Throughout the main trial, make sure that you get participants to keep talking. Otherwise, just give them gentle reminders.

As the initial step in the verbal protocol analysis, the verbal data of two or three participants are randomly selected in order to decide upon a coding scheme for the following analysis. Ryu (2003a) exemplified this process along with evaluating two interactive TV systems.

After transcribing the verbal data sampled (two participants' data), he decided upon a coding scheme containing three categories for handling the mode problems in the interactive TV: *trial and error, recognition of mode signal,* and *recall of interaction history*. In the first category, the participants did not know what action they had to take in the moded situation. In the second category, the participants recognized the mode signal given. In the last category, the other participants recalled what they had done to take the correct action. These categories were listed in the first column of Figure 10.3.

The next "Frequency" column gives the number of participants who reported such a protocol corresponding to each category. The last column, which provides typical utterances, illustrates each category. Please note that one asterisk (*) indicates an excerpt from the participants using the original system, and two asterisks (**) indicate an excerpt from those using the redesigned system.

The frequencies in Figure 10.3, as derived from the protocols, appear to have certain differences between the original system and the redesigned one. In particular, the group with the redesigned system made more frequent use of the mode signal to take their next action, whereas the group with the original system relied on their recall or guessing. In this way, the verbal protocol analysis can be used as a pragmatic form of usability testing.

Category	Frequency		Sample excerpts
	Original system	*Redesigned system*	
Trial and error	9	1	* It seems a bit bizarre. Try TV button to get the TV programme, umm, but does not back to.
			** I thought TV one is the most obvious one to press, the other one Web, kind of, back button takes back me to previous Web site than TV. The green TV one should take me automatically back to TV.
Recognition of mode signal	0	11	** There is no blinking one, it may indicate the different one.
			** I am going back to TV, because the flashing TV screen which means you can go to TV directly.
Recall of interaction history	4	1	* I need to go along with the back button, well, because last time when I typed the URL box I had to go to the back button to get back to TV.
			** Probably use the back button because last time I changed the address I wanted to reach.

Figure 10.3. A verbal protocol analysis, reprinted from Ryu (2003a).

A Guide for Practitioners: For Validity of Verbal Protocol Analysis

The verbal protocol analysis is only taking the analyst's expertise on trust, and to produce valid interpretations one should be able to demonstrate that appropriate precautions had been taken to minimize the problem of both experimenter and participant biases.

Internal validity is concerned with factors within the study itself that could be leading you to wrongly interpret your results in favour of the initial hypothesis you came up with. In relation to this internal validity, the coding scheme applied in the analysis should be demonstrated as the degree to which the coding scheme accurately measures what it is intended to measure.

Also, more than one analyst can independently interpret the same verbal data, and then compare them for the external validity which is concerned with whether or not you can generalize the findings from your study to the way the population at large would behave.

Flow Experience

The concept of *flow experience* (Csikszentmihalyi, 1990) has been highly appropriated by Hoffman and Novak (1996) as a tool for understanding user behaviour—in particular, user experience—to evaluate data such as how satisfied the user is with various aspects of the system given. In spite of the high endorsement by researchers (e.g., Webster and Martocchio, 1992) who have noted that *flow* is a useful construct for describing more pleasant human-computer interaction, little has been discussed as to how to measure *flow experience* in mobile UI testing.

In this respect, this section details and reviews its definition and how to measure the flow experience in terms of respondents' *controls* and *challenges*[77] when it comes to usability testing of mobile systems.

Though there would be many different definitions of "flow," it is generally said that flow is a holistic control sensation where one acts with total involvement, with a narrowing of *focus of attention* (Csikszentmihalyi, 1990). To our interest, it implies that, in order to experience flow while engaged in an activity of using a mobile device, users must perceive a balance between their *controls* and the *challenges* of the activity, and both must be above a critical threshold, which presents users with *playful interaction, exploratory behaviour* and *positive subjective experience* (Hoffman and Novak, 1996).

Indeed, there have been many attempts, such as both focus groups and questionnaires, that aim to capture qualitative aspects of the user interface, e.g., enjoyment or pleasant-to-use, and applied to measure flow. Most notably, Microsoft™ does extensive usability testing, mostly via focus groups, and measures enjoyment by interrupting the user's interaction every few minutes via a dialogue box asking them for their current level of engagement. They rely on quantitative information to firstly identify usability problems, and then go to the richer qualitative data to explain why the problem occurs. Yet, collecting data with this elaborated mechanism is a challenge for evaluating mobile user interfaces. For usability practitioners, building upon Webster, Trevino and Ryan's (1993) study, the following twelve questions would indicate the level of user experience as they are given after selected activities with a mobile device to later evaluate their experience. A good feature of the *performing activity and then filling in survey* method is that it can be used in laboratory usability testing (Ghani, Supnick, and Rooney, 1991) as well.

- *Control*[78] – For an activity to encourage playful, exploratory behaviors, individuals should experience feelings of control over interaction experience given in usability tasks.
    - *Question 1: When using the mobile device, I felt in control.*
    - *Question 2: I felt that I had no control over my interaction with the mobile device.*
    - *Question 3: The mobile device allowed me to control the interaction*

- *Attention focus* – Centering attention is a necessary condition for achieving flow experience, being narrowed to a limited stimulus and cancelling out irrelevant thoughts and perception (Hoffman and Novak, 1996).
    - *Question 4: When using the mobile device, I thought about other things.*
    - *Question 5: When using the mobile device, I was aware of distractions.*
    - *Question 6: When using the mobile device, I was totally absorbed in what I was doing.*

---

[77] These are keys constructs to understanding flow.
[78] Csikszentmihalyi (1990, p.3) pointed out that "we feel in control of our actions and master of our own fate...we feel a sense of exhilaration, a deep sense of enjoyment."

- *Curiosity* – Cognitive curiosity and the desire to attain competence with the technology or the software may motivate the user to learn more skills where the challenges are matched by the user's skills.
    - Question 7: *Using the mobile device excited my curiosity.*
    - Question 8: *Interacting with the mobile device made me curious.*
    - Question 9: *Using the mobile device aroused my imagination.*

- *Playfulness* – Webster, Trevino and Ryan (1993) termed "playfulness" as subjective experiences during interactions that are characterized by perceptions of pleasure and involvement. In the playful state, people are so intensively involved in an activity that nothing else seems to matter (Csikszentmihalyi, 1990), so involvement in a playful, exploratory experience is self-motivating because it is pleasurable and encourages repetition.
    - Question 10: *Using the mobile device bored me.*
    - Question 11: *Using the mobile device was intrinsically interesting.*
    - Question 12: *The mobile device was fun for me to use.*

The flow experience is not good in an absolute sense, so it may be associated with the *Likert scale* or *semantic differential scale* in questionnaire design (see Chapter 4 for further detail). Please note that we, as mobile interface designers, must distinguish between useful and harmful forms of flow, making the most of the former and limiting the latter in interface design, by which the mobile system has the potential to make mobile activities richer, more intense and more meaningful (Csikszentmihalyi, 1990).

Epilogue

THE WAY FORWARD

Mobile user interface analysis and design, as the main theme of this book, is a relatively new research area and more demanding than desktop computer interface design. Hence, there is an ever-increasing desire for appropriate tools and techniques, but perhaps less enthusiasm or support for mobile interface design practitioners to have the opportunity to fully articulate the relationships among these tools, techniques and underlying HCI theories.

In this light, a necessity is a comprehensive structure covering current trends, technologies and techniques in mobile user interface analysis and design. It is my personal belief that this book will be a timely publication for both academics and practitioners who are interested in the design and development of future mobile user interfaces. Obviously, I have arranged the related materials in an order that makes sense to me, trying wherever possible to locate readings that speak to the same or closely-related issues, but many different arrangements are possible, and these reinterpretations may suggest other solutions or design processes to the future challenges of mobile user interface design.

The main objective that I set out in the beginning of this book was to return to the fundamental area of mobile user interface analysis and design, and to prepare you (mobile interaction designers or practitioners) for more advanced designing of the future mobile user interface. However, this book is not all about mobile user interface design. Even though it is still frequently applied to the domain of mobile user interface, this kind of thinking and design process can lead to a truckload of other design achievements I have not mentioned, if and only if all of them concentrate on ensuring that the simplified four-staged user-centered design process remains front and centre.

There are a few dozen books on mobile interface design. They tell a lot about human psychology, how to study users and their tasks, how to test a prototype or finish a system for usability and many other good things. Amazingly, they say very little about the actual design of real-life mobile user interfaces, how we can design screens, what they should contain, how we present data, and how we can make sure that our prototypes can be close to a final good solution.

To be fair, I have seen a few books that explain mobile human-computer interaction: Ballard (2007) and Jones and Marsden (2006)[79], for instance, but not for mobile user interface design itself. Many mobile practitioners have looked into the books about mobile HCI. They read and read, but find little they can use right away. And they do not get a reasonable answer

---

[79] Here, I acknowledge the two books for their valuable insights and studies, which galvanized this book.

to how to design a mobile user interface. The books somehow assume that it is a trial and error process, but we mobile interface designers do not think so.

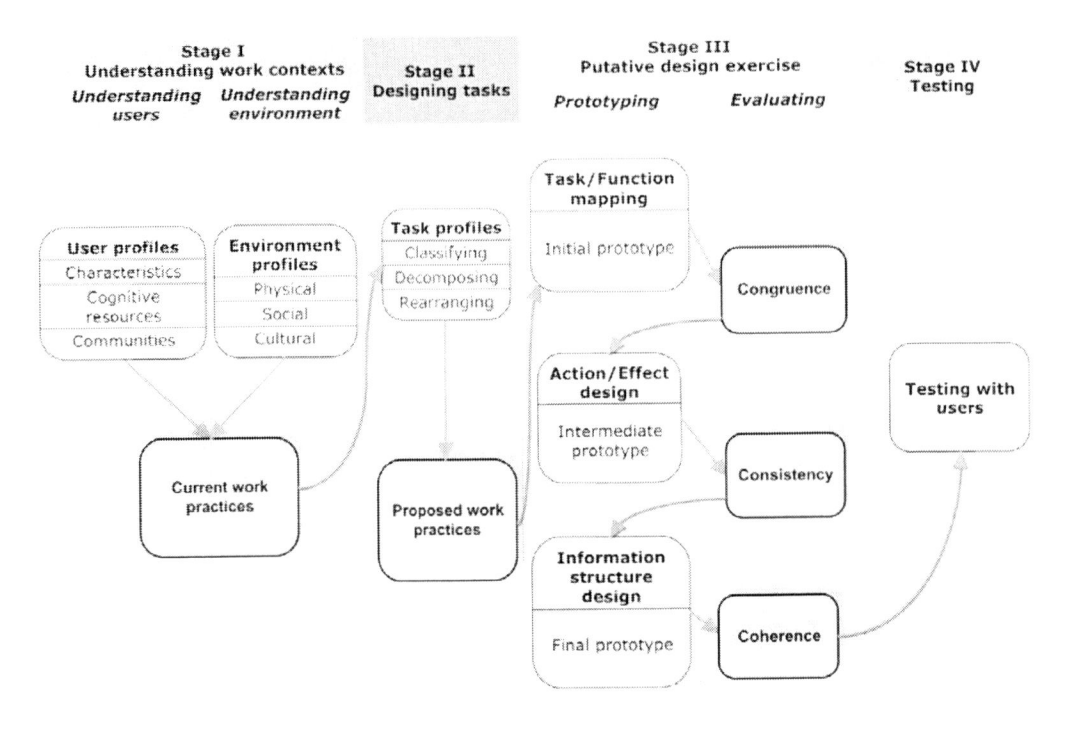

Figure 11.1. The mobile user interface design process.

Hence, what I have tried in this book is to bridge the gap. I have developed the putative interface design process, as shown in Figure 11.1, over my many years of working experience with mobile practitioners and postgraduate students, and it has become a routine exercise to develop a mobile user interface.

Yet, it is not true that if you follow my design process the result is free of usability problems. Instead, modestly speaking, if you follow the design process, your mobile user interface would become right, and the usability problems can be fixed in a much better way than conventional trial-and-error.

As a concluding remark, this book has paid little attention to aesthetics – how to make the mobile devices look pretty and attractive – or to the entertainment value of the mobile device. These issues are very important in some mobile devices, such as mobile phones or portable game consoles (e.g., Nintendo™ DS), but fairly speaking, I am not the expert of this research agenda—indeed, I am very ignorant of this fascinating topic. Further, I have not figured out how to deal with aesthetics in a systematic way. So I wish that my humble thoughts in this book would attract some of the future researchers in the field of mobile user interface design, being conscious of the distinction between mine and the missing parts in mobile user interface analysis and design.

REFERENCES

Alexander, I., and Maiden, N. (Eds.). (2004). *Scenarios, Stories, User Cases: Through the Systems Development Life-Cycle*. New York: Wiley.

Anderson, J. R. (2000). *Cognitive Psychology and Its Implications* (5th ed.). New York: Worth.

Andre, T. S. (2001). The user action framework: A reliable foundation for usability engineering support tools. *International Journal of Human-Computer Studies*, 54(1), 107-136.

Annett, J. (2004). Hierarchical task analysis. In D. Diaper and N. Stanton (Eds.), *The Handbook of Task Analysis for Human-Computer Interaction* (pp. 67-82). Mahwah, NJ: Lawrence Erlbaum.

Apple computer. (1992). *Macintosh Human Interface Guidelines*. Reading, MA: Addison-Wesley.

AvantGo Mobile LifeStyle Survey. (2004). AvantGo mobile lifestyle survey reveals the top twelve "Dream" features of an all-In-one handheld device. Retrieved 09.01, 2008, from http://www.ianywhere.com/press_releases/avantgo_dreams_survey.html

Aykin, N. (Ed.). (2004). *Usability and Internationalization of Information Technology*. New York: CRC Press.

Baddeley, A. (1986). *Working Memory*. Oxford, U.K.: Clarendon.

Bailey, R. W. (2001). Heuristic evaluation vs. user testing. UI design Update Newsletter Retrieved 08.05, 2005, from http://www.humanfactors.com/library/jan001.htm

Balattner, M. M., Sumikawa, D. A., and Greenberg, R. M. (1989). Earcons and icons: Their structure and common design principles. *Human-Computer Interaction*, 4(1), 11-44.

Ballard, B. (2007). *Designing the Mobile User Experience*. Chicester, U.K.: John Wiley and Sons. .

Barber , C. (2001). *An Interactive Heuristic Evaluation Toolkit*. University of Sussex, Sussex, U.K.

Beagley, N. I., Haslam, R. A., and Parsons, K. C. (1993). Hypermedia for the in-house development of information systems. In G. Salvendy and M. Smith (Eds.), *Human-Computer Interaction: Applications and Case Studies*. Amsterdam, The Netherlands: Elsevier.

Beato, G. (2005). Twenty-five years of post-it notes. Retrieved 14.09, 2005, from http://rakemag.com/features/detail.asp?catID=61anditemID=20620

Beck, K. (1999). *Extreme Programming Explained: Embrace Change*. New York: Addison-Wesley.

Bendy, G. Z., Seglin, M. H., and Meister, D. (2000). Activity theory: history, research and application. *Theoretical Issues in Ergonomic Sciences*, 1(2), 168-206.

Benyon, D., Gree, T., and Bental, D. (1998). *Conceptual Modeling for User Interface Development*. London, U.K: Springer-Verlag.

Bertelsen, O. W. (2004). *The activity walkthrough: an expert review method based on activity theory*. Paper presented at the NordiCHI, Tampere, Finland.

Bertelsen, O. W., and Bodker, S. (2003). Activity theory. In J. M. Carroll (Ed.), *HCI Models, Theories and Frameworks: Towards a Multidisciplinary Science* (pp. 291-324). San Francisco, CA: Morgan Kaufman.

Beyer, H., and Holtzblatt, K. (1998). *Contextual Design: Defining Customer-Centred Systems*. San Francisco, CA: Morgan Kaufmann.

Blackmon, M. H., Kitajima, M., and Polson, P. (2003). Repairing usability problems identified by the cognitive walkthrough for the Web. *Paper presented at the Conference on Human Factors in Computing Systems*, Ft. Lauderdale, FL.

Bosma, H., Boxtel, M., Ponds, R., Houx, P., and Jolles, J. (2003). Education and age related cognitive decline: The contribution of mental workload. *Educational Gerontology*, 29(2), 165-173.

Brewster, S. (1997). *Navigating telephone-based interface with earcons*. Paper presented at the BCS-HCI, Bristol, U.K.

Brynat, J. (2003). Map data on the move [Electronic Version]. Retrieved 23.06.2008, from http://www.ordnancesurvey.co.uk/oswebsite/business/sectors/wireless/news/articles/pdf/Map%20data%20on%20the%20move.pdf

Bushey, P. R., Deelman, T., and Mauney, J. M. (2006). USA Patent No.: U. P. office.

Byrne, M. D. (1995). *A working memory model of a common procedural error* (No. GIT-CS-95/06). Atlanta, GA: Georgia Institute of Technology. Document Number).

Card, S. K., Moran, T. P., and Newell, A. (1983). *The Psychology of Human-Computer Interaction*. Hillsdale, NJ: Lawrence Erlbaum.

Carlos, J., Costa, J. S., and Aparicio, M. (2007). *Evaluating web usability using small display devices*. Paper presented at the ACM International Conference on Design of Communication, El Paso, TX.

Carroll, J. (1993). *Human Cognitive Abilities: A Survey of Factor-Analytic Studies*. New York: Cambridge University Press

Carroll, J. (Ed.). (2002). *Human-Computer Interaction in the New Millennium*. New York: ACM Press.

Carroll, J. (Ed.). (2003). *HCI Models, Theories, and Frameworks: Toward a Multidisciplinary Science*. Amsterdam, The Netherlands: Morgan Kaufmann.

Carroll, J., and Carrithers, C. (1984). *Training wheels in a user interface. Communications of the ACM*, 27(8), 800-806.

Carroll, J., Kellogg, W. A., and Rosson, M. B. (1991). The task-artifact cycle. In J. M. Carroll (Ed.), *Designing Interaction: Psychology at the Human-Computer Interface* (pp. 74-102). Cambridge, U.K.: Cambridge University Press.

Carroll, J., and Rosson, M. (1985). Usability specifications as a tool in iterative development. In H. Hartson (Ed.), *Advances in Human-Computer Interaction* (pp. 1-28). Norwoord, NJ: Ablex.

Courage, C., and Baxter, K. (2005). *Understanding Your Users: A Practical Guide to User Requirements - Methods, Tools and Techniques*. San Francisco, CA: Elsevier.

Csikszentmihalyi, M. (1990). *Flow: The Psychology of Optimal Experience*. New York: Harper Perennial.

References

Cuomo, D. L., and Bowen, C. D. (1992). Stages of user activity model as a basis for user-system interface evaluation. Paper presented at the Human Factors and Ergonomics Society Annual Meeting Atlanta, GA.

Davis, F. (1989). Perceived usefulness, perceived ease of use, and user acceptance of information technology. *MIS Quarterly*, 13(3), 319-340.

Dewar, R. (1993). Warning: Hazardous road signs ahead. *Ergonomics in Design*, 1(3), 26-31.

Dewsbery, V. (2005). *Designing for Small Screens* (V. Dewsbery, Trans.). Geneva, Switzerland: AVA Publishing.

Dix, A. J. (2001). Excel mode error. Retrieved 11, 05, 2004, from http://www.comp.lancs.ac.uk/computing/users/dixa/casestudy/excel-mode/

Djajadiningrat, T., Overbeeke, K., and Wensveen, S. (2002). But how, Donald, tell us how? on the creation of meaning in interaction design through feedforward and inherent feedback. *Paper presented at the International Conference on Designing Interactive Systems*, London, U.K.

Donk, M. (1994). Human monitoring behaviour in a multiple-instrument setting: Independent sampling, sequential sampling or arrangement-dependent sampling. *Acata Psychologia* 86(1), 31-55.

Dourish, P. (2001). *Where the Action Is: The Foundations of Embodied Interaction*. Cambridge, MA: MIT Press.

Draper, S. W., and Barton, S. B. (1993). Learning by exploration, and affordance bugs. *Paper presented at the Conference on Human Factors in Computer Systems*, Amsterdam, The Netherlands.

Drury, C. (1983). Task analysis methods in industry. *Applied Ergonomics*, 14(1), 19-28.

Duke, D. J., Barnard, P. J., Duce, D. A., and May, J. (1998). Syndetic modelling. *Human-Computer Interaction*, 13(4), 337.

Duncan, K. D. (1972). Strategies for the analysis of the task. In J. Hartley (Ed.), *Programmed Instruction: An Education Technology* (pp. 19-81). London, U.K: Butterworths.

Ehn, P., and Kyng, M. (1991). *Cardboard Computers: Mocking-it-up or Hands-on the Future*. In J. Greenbaum and M. Kyng (Eds.), Design at Work (pp. 169-196). Hillsdale, NJ: Laurence Erlbaum

Engeström, Y., and Middleton, D. (1996). *Cognition and Communication at Work*. Boston, MA.: Cambridge University Press.

Ericsson, K. A., and Simon, H. A. (1985). *Protocol Analysis: Verbal Reports as Data*. Cambridge, MA.: MIT Press.

Esbjörnsson, M., Juhlin, O., and Weilenmann, A. (2007). Drivers using mobile phones in traffic: An ethnographic study of interactional adaptation. *International Journal of Human-Comptuer Interaction*, 22(1), 39-60.

Faulkner, L. (2003). Beyond the five-user assumptions: Benefits of increased sample sizes in usability testing. *Behavior Research Methods, Instruments and Computers*, 35(3), 379-383.

Fisher, R. (1982). *Social Psychology: An Applied Approach*. New York: St. Martin's Press.

Flanagan, J. C. (1954). The critical incident technique. *Psychological Bulletin*, 51(4), 327-358.

Fortunati, L. (2002). Italy: Stereotypes, true and false. In J. E. Katz and M. Askhus (Eds.), *Perpetual Contact: Mobile Communication, Private Talk, Public Performance* (pp. 42-62). Cambridge, U.K.: Cambridge University Press.

Frese, M., and Zapf, D. (1994). Action as the core of work psychology: A German approach. In H. C. Triandis, M. D. Dunnette and L. M. Hougheds (Eds.), *Handbook of Industrial and Organizational Psychology* (2nd ed., Vol. 4, pp. 271-340). Palo Alto, CA: Consulting Psychologists.

Freyd, J. J., and Finke, R. A. (1984). Representational momentum. *Journal of Experimental Psychology: Learning, Memory, and Cognition*, 10, 126-132.

Gamble, R. (1986). Cognitive momentum. *Physics Education*, 21(1), 24-27.

Gaver, B., Dunne, T., and Pacenti, E. (1999). Cultural probes. *ACM Interactions*, 6, 21-29

Ghani, J. A., Supnick, R., and Rooney, P. (1991). The experience of flow in computer-mediated and in face-to-face groups. *Paper presented at the International Conference on Information Systems*, New York.

Gibson, J. J. (1979). *The Ecological Approach to Visual Perception*. Boston, MA: Houghton-Mifflin.

Goodman, J., Brewster, S., and Gray, P. (2004). Older people, mobile devices and navigation. *Paper presented at the Workshop on HCI and Older Population in the BCS HCI 2004*, Leeds, U.K.

Gryazin, E. A., Tuominen, J. O., and Kourouthanassis, P. (2003). Distributed information systems development for consumer market. *Paper presented at the IEEE International Conference Artificial Intelligent Systems*, Divnomorskoe, Russia.

Hastie, T., Tibshirani, R., and Friedman, J. H. (2003). *The Elements of Statistical Learning* New York: Springer.

Hertzum, M., and Jacobsen, N. E. (2001). The evaluator effect: A chilling fact about usability evaluation methods. *International Journal of Human-Computer Interaction*, 13(4), 421-443.

Hinckley, K., and Sinclair, M. (1999). Touch-sensing input devices. *Paper presented at the Conference on Human Factors in Computing Systems*, Pittsburgh, PA.

Hoffman, D. L., and Novak, T. P. (1996). Marketing in hypermedia computer-mediated environments: Conceptual foundations. *Journal of Marketing*, 60(July), 50-68.

Hughes, J., King, V., Rodden, T., and Andersen, H. (1995). The role of ethnography in interactive systems design. *ACM Interactions*, 2, 56-65.

Jacobsen, N. E., and John, B. E. (2000). *Two case studies in using cognitive walkthrough for interface evaluation* (No. CMU-CS-00-132). Pittsburgh, PA: Computer Science Department, Carnegie-Mellon University. Document Number)

Janassen, D. H., Tessmer, M., and Hannum, W. H. (1999). *Task Analysis Methods for Instructional Design. Mahwah*, NJ.: Lawrence Erlbaum.

Jones, M., and Marsden, G. (2006). *Mobile Interaction Design*. Chichester, U.K.: John Wiley.

Jones, R. (2008). Ethnography: Adding reality and penetrating insight, or past its heyday? *Interfaces*, 74, 12-14.

Jones, S., and Burnett, G. E. (2007). Children's navigation of hyperspace: Are spatial skills important? . *Paper presented at the Conference on IASTED International Conference Web-Based Education*, Chamonix, France.

Kaikkonen, A., Kekäläinen, A., Cankar, M., Kallio, T., and Kankainen, A. (2005). Usability testing of mobile applications: A comparison between laboratory and field testing. *Journal of Usability Studies*, 1(1), 4-16.

Karat, C.-M. (1994). A comparison of user interface evaluation methods. In J. Nielsen and R. L. Mack (Eds.), *Usability Inspection Methods* (pp. 203-233). New York: Wiley.

Karat, C.-M., Campbell, R., and Fiegel, T. (1992). Comparison of empirical testing and walkthrough methods in user interface evaluation. *Paper presented at the Conference on Human Factors in Computing Systems*, Monterey, CA.

Kasesniemi, E., and Rautiainen, P. (2002). Mobile culture of children and teenagers in Finland. In J. E. Katz and M. Aakhus (Eds.), *Perpetual Contact: Mobile Communication, Private Talk, Public Performance*. Cambridge, U.K.: Cambridge University Press.

Katz, J. E., and Aakhus, M. (Eds.). (2002). *Perpetual contact: Mobile communication, private talk, public performance*. Cambridge, U.K: Cambridge University Press.

Kazdin, A. E. (2003). *Research Design in Clinical Psychology*. Boston, MA: Allyn and Bacon.

Kelley, J. F. (1984). An iterative design methodology for user-friendly natural language office information applications. *ACM Transactions on Office Information Systems*, 2(1), 26-41.

Kieras, D. E., and Polson, P. G. (1985). An approach to the formal analysis of user complexity. *International Journal of Man-Machine Studies*, 22(4), 365-394.

Kim, J. H., and Lee, K. P. (2005). Cultural difference and mobile phone interface design: Icon recognition according to level of abstraction. *Paper presented at the International Conference on Human-Computer Interaction with Mobile Devices and Services*, Salzburg, Austria.

Kirwan, B., and Ainsworth, L. (1992). *A Guide to Task Analysis*. London, U.K: Taylor and Francis.

Knight, J. (2008). Case study: Guerrilla interaction design of an Intranet. *Interfaces*, 75, 10-12.

Krueger, R. A. (1994). *Focus Groups: A Practical Guide for Applied Research*. London, U.K.: Sage Publications.

Kules, B., Kustanowitz, J., and Shneiderman, B. (2006). Categorizing web search results into meaningful and stable categories using fast-feature technique. *Paper presented at the ACM/IEEE-CS Joint Conference on Digital Libraries Chapel Hill*, NC.

Kurosu, M., and Kashimura, K. (1995). Apparent usability vs. inherent usability: Experimental analysis on the determinants of the apparent usability. *Paper presented at the Conference on Human Factors in Computing Systems,* Denver, CL.

Kutti, K. (1996). Activity theory as a potential framework for human-computer interaction research. In B. A. Nardi (Ed.), *Context and Consciousness: Activity Theory and Human-Computer Interaction*. Cambridge, MA: MIT Press.

Lee, W. H., and Ryu, H. (2007). Design guidelines for map-based human-robot interface: A co-located workspace perspective. *International Journal of Industrial Ergonomics*, 37(7), 589-604.

Leplatre, G., and Brewster, S. (2000). Designing non-speech sounds to support navigation in mobile phone menus. *Paper presented at the International Conference on Auditory Display*, Atlanta, GA.

Lewis, C., and Wharton, C. (1997). Cognitive walkthroughs. In M. Helander, T. K. Landauer and P. Prabhu (Eds.), *Handbook of Human-Computer Interaction*. Amsterdam, The Netherlands: Elsevier.

Ling, R. (2004). *The Mobile Connection: The Cell Phone's Impact on Society*. San Francisco, CA: Morgan Kaufmann.

Ling, R., and Yttri, B. (2002). Hyper-coordination via mobile phones in Norway. In J. E. Katz and M. Aakhus (Eds.), *Perpetual Contact: Mobile Communication, Private Talk, Public Performance* (pp. 139-169). Cambridge, U.K.: Cambridge University Press.

Liu, L., and Khooshabeh, P. (2003, April 5 -10). Paper or interactive? A study of prototyping techniques for ubiquitous computing environments. *Paper presented at the Conference on Human Factors in Computing Systems*, Ft. Lauderdale, FL.

Love, S., Foster, J. C., and Jack, M. A. (1997). Assaying and isolating individual differences in automated telephone services. *Paper presented at the International Conference on Human Factors in Telecommunications*, Oslo, Norway.

Love, S., Foster, J. C., and Jack, M. A. (2000). Health warning: Use of speech synthesis can cause personality change. *Paper presented at the State of the Art in Speech Synthesis*, London, U.K.

Luchini, K., Quintana, C., and Soloway, E. (2004, 24-29 April). Design guidelines for learner-centered handheld tools. *Paper presented at the Conference on Human Factors in Computer Systems*, Vienna, Austria.

Lyons, K., Starner, T., Plaisted, D., Fusia, J., Lyons, A., and Drew, A. (2003). Twiddler typing: one-handed chording text entry for mobile phones. *Paper presented at the Conference on Human Factors in Computing Systems*, Ft. Lauderdale, FL.

Mack, R. L., and Montaniz, F. (1994). Usability inspection methods. In J. Nielsen and R. L. Macks (Eds.), *Usability Inspection Methods* (pp. 295-339). New York: John Wiley.

MacKenzie, I. S., and Soukoreff, R. W. (2002). Text entry for mobile computing: models and methods, theory and practice. *Human-Computer Interaction*, 17(2), 147-198.

MacKenzie, I. S., and Zhang, S. X. (1999). The design and evaluation of a high-performance soft keyboard. *Paper presented at the Conference on Human Factors in Computing Systems*, Pittsburgh, PA

MacLean, A., Young, R. M., Bellotti, V. M. E., and Moran, T. (1996). Questions, options, and criteria: Elements of design space analysis. In T. Moran and J. Carroll (Eds.), *Design Rationale Concepts, Techniques, and Use* (pp. 53-106). New York: Lawrence Erlbaum Associates.

Marila, J., and Ronkainen, S. (2004). Time-out in user interface: the case of mobile text input. *Personal and Ubiquitous Computing*, 8(2), 110-116.

Maurer, D., and Warfel, T. (2008). Card sorting: A definitive guide [Electronic Version]. Boxes And Arrows: The Design Behind the Design, from http://www.boxesandarrows.com/view/card_sorting_a_definitive_guide

Mayhew, D. (1999). *The Usability Engineering Lifecycle: A Practitioner's Handbook for User Interface Design*. San Francisco, CA: Morgan Kaufmann.

McDonald, J. E., Stone, J. D., and Liebelt, L. S. (1983). Searching for items in menus: The effects of organization and type of target. *Paper presented at the Human Factors and Ergonomics Society Annual Meeting Santa Monica*, CA.

McKinney, V., Wilson, D., Brooks, N., O'Leary-Kelly, A., and Hardagrve, B. (2008). Women and men in the IT profession. *Communications of the ACM*, 51(2), 81-84.

Microsoft. (1995). *The Windows Interface Guidelines for Software Design: An Application Design Guide*. Seattle, WA: Microsoft Press.

Moggridge, B. (Ed.). (2007). *Designing Interactions*. Cambridge, MA: MIT Press.

Monk, A. (1986). Mode errors - A user-centered analysis and some preventative measures using keying-contingent sound. *International Journal of Man-Machine Studies*, 24(4), 313-327.

Monk, A. (1990). Action-effect rules - a technique for evaluating an informal specification against principles. *Behaviour and Information Technology*, 9(2), 147-155.

Monk, A. (1997). Lightweight techniques to encourage innovative user interface design. In L. Wood and R. Zeno (Eds.), *Bridging the Gap: Transforming User Requirements into User Interface Design*. Boca Raton, FL: CRC Press.

Monk, A. (1998). Cyclic interaction: a unitary approach to intention, action and the environment. *Cognition*, 68(2), 95-110.

Monk, A. (1999). *Modelling cyclic interaction. Behaviour and Information Technology*, 18(2), 127-139.

Monk, A. (2003). Common ground in electronically mediated communication: Clark's theory of language use In J. Carroll (Ed.), *HCI Models, Theories, and Frameworks: Toward a Multidisciplinary Science*. Amsterdam, The Netherlands: Morgan Kaufmann.

Monk, A., Carroll, J., Parker, S., and Blythe, M. (2004). Why are mobile phones annoying? *Behaviour and Information Technology*, 23(1), 33-41.

Monk, A., and Howard, S. (1998). The rich picture: A tool for reasoning about work context. *ACM Interactions*, 5, 21-30.

Monk, A., Nardi, B. A., Gilber, N., Mantei, M., and McCarthy, J. (1993). Mixing oil and water?: Ethnography versus experimental psychology in the study of computer-mediated communicaton. *Paper presented at the Conference on Human Factors in Computing Systems*, Amsterdam, The Netherlands.

Monk, A., Wright, P., Haber, J., and Davenport, L. (1993). *Improving Your Human-Computer Interface:* A Practical Technique New York: Prentice Hall.

Moran, T. P. (1983). Getting into a system: External-Internal task mapping analysis. *Paper presented at the Conference on Human Factors in Computing Systems*, Boston, MA.

Moray, N. (1981). The role of attention in the detection of errors and the diagnosis of errors in man-machine system failures. In J. Rasmussen and W. Rouse (Eds.), *Human Detection and Diagnosis of System Failures*. New York: Plenum Press.

Moray, N. (1986). Monitoring behaviour and supervisory control. In K. R. Boff, L. Kaufman and J. P. Thomas (Eds.), *Handbook of Perception and Human Performance*. New York: Wiley.

Morris, M., and Venkatesh, V. (2000). Age differences in technology adoption decisions: Implications for a changing work force. *Personnel Psychology*, 53(2), 375-403.

Morrison, J. B., and Tversky, B. (2001). The (in)effectiveness of animation in instruction. *Paper presented at the Conference on Human Factors in Computing Systems*, Seattle, WA.

Myers, I. (1993). *Introduction to Type*. Palo Alto, CA: Consulting Psychologists Pres.

Nam, T. (2005). Sketch-based rapid prototyping platform for hardware-software integrated interactive products. *Paper presented at the Conference on Human Factors in Computing Systems,* Portland, OR.

Nam, T., and Lee, W. (2003). Integrating hardware and software: augmented reality based prototyping method for digital products. *Paper presented at the Conference on Human Factors in Computing Systems*, Ft. Lauderdale, FL.

Nardi, B. A. (Ed.). (1997). *Context and Consciousness: Activity Theory and Human-Computer Interaction.* Cambridge, MA: MIT Press.

Nehrling, M. (2005). Mobile learning design - The post-it method. M-Learning World Retrieved 14.09, 2005, from http://www.mlearningworld.com/index.php?name=Articlesandop=showandid=1453.

Newell, A. (1990). *Unified Theories of Cognition.* Cambridge, MA: Harvard University Press.

Nielsen, J. (1994). Heuristic evaluation. In J. Nielsen and R. L. Mack (Eds.), *Usability Inspection Methods.* New York: Wiley.

Norman, D. A. (1988). *The Psychology of Everyday Things.* New York: Basic Books.

Norman, D. A. (1991). *The Psychology of Menu Selection: Designing Cognitive Control at the Human/Computer Interface.* Norwood, NJ: Ablex.

Norman, D. A. (1999). Affordance, conventions, and design. *ACM Interactions,* 6, 38-42.

Norman, D. A. (2004). *Emotional Design.* New York: Basic Books.

Ouden, P. H. d. (2006). *Development of a Design Analysis Model for Consumer Complaints: Revealing a New Class of Quality Failures.* TUE, Eindhoven, The Netherlands.

Palanque, P., and Bastide, R. (1996). Task models - system models: a formal bridge over the gap. In D. Benyon and P. Palanque (Eds.), *Critical Issues in User Interface System Engineering:* Springer-Verlag.

Park, J., Yoon, W. C., and Ryu, H. (2000). Users' recognition of semantic affinity among tasks and the effects of consistency. *International Journal of Human-Computer Interaction,* 12(1), 89-105.

Parsons, D., Ryu, H., Lal, R., and Ford, S. (2005). Paper prototyping in a design framework for professional mobile learning. *Paper presented at the International We-B (Working for E-Business) Conference Sydney,* Australia

Payne, S. J., and Green, T. R. G. (1986). Task-action grammars: a model of the mental representation of task languages. *Human-Computer Interaction,* 2(2), 93-133.

Pirhonen, A., Brewster, S., and Holguin, C. (2002). Gestural and audio metaphors as a means of control for mobile devices. *Paper presented at the Conference on Human Factors in Computing Systems,* Minneapolis, MN.

Polson, P. G., Lewis, C., Rieman, J., and Wharton, C. (1992). Cognitive walkthroughs - A method for theory-based evaluation of user interfaces. *International Journal of Man-Machine Studies,* 36(5), 741-773.

Polson, P. G., and Lewis, C. H. (1990). Theory-based design for easily learned interfaces. *Human-Computer Interaction,* 5(2), 191-220.

Posner, M. L., Nissen, J. M., and Klein, R. (1976). Visual dominance: An information processing account of its origins and significance. *Psychological Review,* 83(2), 157-171.

Preece, J., Yvonne, R., and Helen, S. (2007). *Interaction Design* (2nd ed.). New York: John Wiley and Sons.

Rasmussen, J. (1986). *Information Processing and Human-Machine Interaction: An Approach to Cognitive Engineering.* New York: North-Holland.

Reisner, P. (1993). APT - a description of user-interface inconsistency. *International Journal of Man-Machine Studies,* 39(2), 215-236.

Roger, Y. (1989). Icons at the interface: Their usefulness. *Interacting with Computers,* 1(1), 105-117.

Ryu, H. (2003a). *A Framework for Interaction Design.* Unpublished Doctoral thesis, University of York, York, U.K.

Ryu, H. (2003b). Modelling cyclic interaction: an account of goal-elimination process. *Paper presented at the Conference on Human Factors in Computing Systems*, Ft. Lauderdale, FL.

Ryu, H. (2006). Noddy's guide to mode problems. *Interfaces*, 66, 10-13.

Ryu, H., and Monk, A. (2004a). A brief account of interaction problems. *Paper presented at the Australian Conference on Human Factors of Computing Systems*, Wollongong, Australia.

Ryu, H., and Monk, A. (2004b). Analysing interaction problems with cyclic interaction theory: Low-level interaction walkthrough. *PsychNology*, 2(3), 304-330.

Ryu, H., and Monk, A. (2005). Will it be a capital letter: Signalling case mode in mobile devices. *Interacting with Computers*, 17(4), 395-418.

Ryu, H., and Monk, A. (in press). Interaction unit analysis: A new interaction design framework. *Human-Computer Interaction*.

Ryu, H., and Parsons, D. (Eds.). (2008). *Innovative Mobile Learning: Technologies and Techniques.*

Ryu, H., and Wong, A. (2008). Perceived usefulness and performance of human-to-human communications on television. *Computers in Human Behavior*, 24, 1364-1384.

Scialfa, C. T., Kiline, D. W., and Lyman, B. J. (1987). Age differences in target identification as a function of retinal location and noise level: Examination of the useful field of view. *Psychology and Aging*, 2(1), 14-19.

Scornavacca, E., Huff, S., and Marshall, S. (2007). Developing a SMS-based classroom interaction system. Paper presented at the Conference on Mobile Learning Technologies and Applications. from http://molta.massey.ac.nz/massey/depart/sciences/conferences/molta/molta-proceedings.cfm

Shneiderman, B., and Plaisant, C. (2005). *Designing the User Interface: Strategies for Effective Human-Computer Interaction* (Fourth Edition ed.). New York: Addison Wesley.

Silfverberg, M. (2003). The one-row keyboard - A case study in mobile text input. In C. Lindholm, T. Keinonen and H. Kiljander (Eds.), *Mobile Usability: How Nokia Changed the Face of the Mobile Phone* (pp. 157-170). New York: McGraw-Hill.

Silfverberg, M., MacKenzie, I. S., and Korhonen, P. (2000). Predicting text entry speed on mobile phones. *Paper presented at the Conference on Human Factors in Computing Systems*, The Hague, The Netherlands.

Smith, D. C. S., Irby, C., Kimball, R., Verplank, B., and Harlem, E. (1982). Designing the start user interface. *BYTE*, 7(4), 242-282.

Snelders, H. M. J. J., Schoormans, J. P. L., and De Bont, C. J. P. M. (1993). Consumer-product interaction and the validity of conjoint measurement: The relevance of the feel/think dimension. *European Advances in Consumer Research*, 1, 142-147.

Spence, I., and Carey, B. (1991). Customers do not want frozen specifications. *Software Engineering Journal*, 6(4), 175-181.

Spohrer, J. (1999). Information in places. *IBM Systems Journal*, 38(4), 602-628

Stanney, K., and Salvendy, G. (1995). Information visualisation: assisting low spatial individuals with information access tasks through the use of visual mediators *Ergonomics*, 38(6), 1184-1198.

Stolzoff, N., Chuan-Fong Shih, E., and Venkatesh, A. (2000). *The home of the future: An ethnographic study of new information technologies in the home*: University of Californiao. Document Number).

Tanik, M., and Yeh, T. (1989). Rapid prototyping in software development. *IEEE Computer*, 22(5), 9-11.

Tesler, L. (1981). The smalltalk environment. *BYTE*, 6(8), 90-147.

Thelen, E., and Smith, L. (1999). *A Dynamic Systems Approach to the Development of Cognition and Action Cambridge*, MA: MIT Press.

Thompson, D., Hamilton, R., and Rust, T. (2005). Feature fatigue: When product capabilities become too much of a good thing. *Journal of Marketing Research*, 42(November), 431-442.

Tian, Z. Z., Kyte, M. D., and Messer, C. J. (2002). Parallax error in video-image systems. *Journal of Transportation Engineering*, 128(3), 218-223.

Tufte, E. (1990). *Envisioning Information*. Cheshire, CT: Graphics Press.

Uden, L. (2007). Activity theory for designing mobile learning. *International Journal of Mobile Learning and Organisation*, 1(1), 81-102.

Venkatesh, A. (1996). Computers and other interactive technologies for the home. *Communications of the ACM*, 39, 47-54.

Vicente, K. J. (1999). *Cognitive Work Analysis: Towards Safe, Productive and Healthy Computer-Based Work*. Mahwah, NJ.: Lawrence Erlbaum.

Vicente, K. J., Hayes, B. C., and Williges, R. C. (1988). Assaying and isolating individual differences in searching a hierarchical file system. *Human Factors*, 29(3), 349-359.

Vicente, K. J., and Williges, R. C. (1988). Accommodating individual differences in searching a hierarchical file system. *International Journal of Man-Machine Studies*, 29(6), 647-668.

Virzi, R. A. (1992). Refining the test phase of usability evaluation: How many subjects is enough? *Human Factors*, 34(4), 457-468.

Webster, J., and Martocchio, J. J. (1992). Microcomputer playfulness: Development of a measure with workplace implications. *MIS Quarterly*, 16(June), 201-226.

Webster, J., Trevino, L. K., and Ryan, L. (1993). The dimensionality and correlates of flow in human-computer interactions. *Computers in Human Behavior*, 9(4), 411-426.

Wharton, C., Bradford, J., Jeffries, R., and Franzke, M. (1992). Applying cognitive walkthroughs to more complex user interfaces: experiences, issues, and recommendations. *Paper presented at the Conference on Human Factors in Computing Systems*, Monterey, CA.

Wickens, C. D., and Hollands, J. (1999). *Engineering Psychology and Human Performance* (3rd ed.). New York: Pearson.

Wigdor, D., and Balakrishnan, R. (2004). A comparison of consecutive and concurrent input text entry techniques for mobile phones. *Paper presented at the Conference on Human Factors in Computing Systems*, Vienna, Austria.

Wilson, J. R., and Corlett, E. N. (Eds.). (1995). *Evaluation of Human Work*. New York: CRC Press.

Wright, P., Fields, R. E., and Harrison, M. D. (2000). Analyzing human-computer interaction as distributed cognition: the resources model. *Human-Computer Interaction*, 15(1), 1-41.

Wright, P., and Monk, A. (1991). A cost-effective evaluation method for use by designers. *International Journal of Man-Machine Studies*, 35(6), 891-912.

Yang, S. J. H., and Shao, N. W. Y. (2006). An ontology based content model for intelligent Web content access services. *International Journal of Web Service Research*, 3(2), 59-78.

Young, R. M. (1983). Surrogates and mappings: two kinds of conceptual models for interactive devices. In D. Gentner, and Stevens, A.L (Ed.), *Mental models*. Hillsdale, NJ: Lawrence Erlbaum.

Young, R. M., Green, T. R. G., and Simon, T. (1989). Programmable user models for predictive evaluation of interface designs. *Paper presented at the Conference on Human Factors in Computing Systems*, Austin, TX.

Ziefle, M., Scroeder, U., Strenk, J., and Michel, T. (2007). How younger and older adults master the usage of hyperlinks in small screen devices. *Paper presented at the Conference on Human Factors in Computing Systems*, San Jose, CA.

INDEX

3

3G. *See* Third generation

A

Abilities
   Motor, 27, 39, 109, 147, 149, 170
   Spatial, 14, 54, 158, 160, 170
   Verbal, 14, 55
Accelerometer, 143
Acquiescence effect, 76, 227
Action design, 147–50
Action procedure design, 140–42
Action-to-Effect design, 9, 33, 95, 104, 137–53, 153, 166, 167, 177, 178, 179, 181, 190, 202
   Affordance, 104, 109, 110, 112, 114, 115, 116, 121, 123, 128, 140, 176, 178
   Mode design, 142
Action-to-effect problem, 178, 202
Activities
   Cultural, 83
   Cyclic, 35
   Intrinsic, 208, 212
   Mental, 27, 39, 50, 99, 107, 121, 176
   Physical, 27, 99, 177
   Social, 57
Activity checklist, 65
Activity Theory, 29, 61–66, 174, 208
   Activity checklist, 65
   Community, 63, 65, 66
   Division of labor, 63
   Mediating artifacts, 62, 63, 65
   Rules, 62, 63, 65, 66
Activity walkthrough, 208–14
   Cognitive walkthrough, 208
   Interaction unit walkthrough, 208
Adobe Device Central C3, 164
Affordability, 22

Affordance, 104, 109, 110, 112, 114, 115, 116, 121, 123, 128, 140, 176, 178
   Interaction Unit Model, 104, 109, 110, 112, 114, 115, 116
   Physical affordance, 115, 176, 178
   Semantic affordance, 115, 128, 140, 176, 178
   Strong affordance, 182, 183, 187, 202
   Weak affordance, 190
   Wizard of Oz, 222
Age
   Attitude, 39, 57, 66
   Older users, 41, 55, 57, 90, 91
   Technology Adoption, 43
   Useful Field of View (UFOV), 52
   User groups, 41, 44, 68
   Younger users, 41
Agent Partitioning Theory, 190, 191
   Consistency, 190
Always-available connection, 18, 44
Animation, 53, 131
Apple
   iPhone, 5, 8, 15, 36, 48, 52, 123, 143, 147, 148, 163, 164
   iPod, 11, 13, 45, 143
   iPod Touch, 54, 124, 163
   MessagePad, 4
   Mobile data capture system, 21
Application
   Emails, 6
   Fax, 6
   Mobile Business, 22–23
   Mobile Healthcare, 21–22
   Mobile Internet, 6, 8, 19, 105, 106, 147, 194
   Mobile Learning, 19–21, 80, 81, 88, 90, 218
   Office applications, 6
APT. *See* Agent Partitioning Theory
AT. *See* Activity Theory
Attention, 17, 41, 45, 49–53
   Flow experience, 235

Menu design, 50
Selective attention, 50
Attitude, 39, 45, 57, 66, 68, 75, 76, 77, 80, 83
Automatic categorization, 40
Automatic web adaptation, 6, 15

B

Bias, 226–27
   Acquiescence effect, 76, 227
   Experimenter bias, 226
   Hawthorne effect, 227, 229, 231
   Participant bias, 76, 226, 227, 229, 231
Blackberry, 42
Blackberry Syndrome, 18
Bluetooth, 18, 20, 93, 195
BREW, 18
Brief usability testing, 220–23
   Participatory evaluation, 221
   Quick and dirty usability testing, 221
   Talking with existing users, 222
   Wizard of Oz technique, 222
Buyer vs. User, 11

C

Card sorting, 88, 128–29
   Closed card sorting, 128
   Limitations, 129
   Menu design, 158
   Navigation design, 163
   Open card sorting, 128
   Work analysis, 129
Carrier network, 16, 18, 19, 42
Carry principle, 11, 12, 14, 40, 55, 120, 121
CCT. *See* Cognitive Complexity Theory
Ceiling effect, 77
Challenges
   Device proliferation, 18
   Feature fatigue, 1, 12, 13
   Killer applications, 19–25
   Paradox of consumer, 14
   Small screen, 14–17, 18, 25, 52, 54, 55, 131, 151, 154, 155, 159, 160, 163, 229
   Standardization, 17–18
Characteristics
   Cognitive, 9, 45, 69, 155
   Cultural environment, 59
   Desktop, 9, 12, 17
   Environment, 55, 97
   Gender, 44, 45, 67
   Mobile user, 44
   Mobile User, 40
   Physical environment, 56, 66
   Social, 44, 57
   Teens, 44
   User, 30, 35, 39, 40, 44, 68, 90
Chording keyboard, 149
ChordTap, 149
CIM. *See* Critical Incident Method
Closure, 26, 48, 53, 157
Cognition
   Gestalt principles, 47, 53, 155
   Memory, 26, 53–55, 88, 90, 109, 131, 149, 158, 160
   Perception, 46, 49
Cognitive abilities
   Older users, 52
   Younger users, 52
Cognitive characteristics, 45–55
Cognitive Complexity Theory, 174
Cognitive dissonance, 149, 176
Cognitive momentum, 141, 151
Cognitive Psychology, 45
   Information design, 49–53
Cognitive task analysis, 108–14, 114–17
   Decomposition-based task analysis, 98, 99, 102, 106, 116
   user's knowledge, 95, 98, 99, 103, 108, 109, 115
Cognitive walkthrough, 190, 191, 193, 213
   Activity walkthrough, 193
   Interaction unit walkthrough, 193
Cognitive workload, 14, 15, 98, 99, 116, 162, 220, 229
Coherence, 10, 33, 35, 155
Collective walkthrough, 167–215
   Activity walkthrough, 208
   Cognitive walkthrough, 193
   Interaction unit walkthrough, 174
Color coding, 133, 156
Conceptual inquiry, 80
Conceptual model
   Coupling, 119, 124, 126
   Icon, 28, 36, 41, 48, 52, 121, 123, 140, 143, 144, 154, 155, 157, 158, 187
   Metaphor, 41, 123, 128, 131, 143, 154, 157
   Principles, 25, 28, 138, 155, 168, 192, 213, 217
   Task-Action mapping, 126
   Task-Function mapping, 119, 124
Congruence, 10, 32, 35, 119, 124, 126, 174, 175, 176–90, 177
   Interaction unit model, 176
   Interaction unit scenarios, 33, 109, 110, 174, 176
   Interaction unit walkthrough, 176
   Physical affordance, 140, 176, 178
   Semantic affordance, 140, 176, 178
   Task-Function, 128

Consistency, 10, 35, 137, 170–72, 174
   Advantage, 51, 174
   Cross functions, 137
   Design, 51, 108, 124, 128
   Functional commitments consistency, 170
   Mental model, 51, 108, 124, 128
   Procedural consistency, 170
   Three types of consistency, 170, 171, 175
   Visual consistency, 170
Context
   Cultural, 1, 19, 20, 21, 30, 35, 37, 38, 39, 40, 42, 43, 49, 55, 58, 61, 87, 88, 90, 93, 96, 106
   Ethnography, 82
   Observation, 39, 73
   Organizational, 1, 19, 20, 21, 30, 35, 37, 38, 40, 42, 43, 49, 55, 58, 60, 61, 87, 88, 90, 93, 96, 106
   Physical, 1, 19, 20, 21, 30, 35, 37, 38, 39, 40, 42, 43, 49, 55, 58, 60, 61, 87, 88, 90, 93, 96, 106
   social, 1, 19, 20, 21, 30, 35, 37, 38, 39, 40, 42, 43, 49, 55, 58, 60, 61, 82, 87, 88, 90, 93, 96, 106
   Technical, 1, 19, 20, 21, 30, 35, 37, 38, 39, 40, 42, 43, 49, 55, 58, 60, 61, 87, 88, 90, 93, 96, 106
Context awareness, 89
Contextual Design Framework, 26, 33
Continuum of use, 223
Convergent information device, 93
Corporate design patterns, 154, 170, 171
   Apple, 164
Coupling
   Task-Action, 126
   Task-Function, 119, 124
Critical Incident Method, 30, 33, 87–90, 222
   Interview, 30
   Testing, 222
CTA. *See* Cognitive task analysis
   GOMS, 108
   Interaction Unit scenarios, 109
Cultural form factors, 36
Cultural icons, 41, 42
Cultural probe, 86
Cultural requirements, 36, 41, 43, 55, 63, 83, 88
Customization, 40
   Flexibility, 130, 172
   Personalization, 40
Cyclic design, 27, 28, 138, 139, 140, 178
   Cyclic interaction model, 28, 139
   Cyclic interaction theory, 140, 178
Cyclic interaction model, 28
Cyclic interaction theory, 138–40, 138, 140, 178

D

Data gathering
   Ceiling effect, 77
   Critical Incident Method, 87–90
   Diary studies, 84–86
   Ethnography, 82–87
   Floor effect, 77
   Focus group, 81–82
   Interview, 78–81
   Observation, 73–75
   Questionnaire, 75–78
   Think-aloud, 217, 220, 231–34, 232
Design
   Action design, 147
   Action procedure design, 140
   Action-to-Effect, 9, 33, 95, 104, 137, 153, 166, 167, 177, 178, 179, 181, 190
   Activity Theory, 61–66
   Affordance, 104, 109, 110, 112, 114, 115, 116, 121, 123, 128, 140, 176, 178
   Carroll and Rosson's view, 3
   Coherence, 10, 33, 35, 155
   Complexity, 1, 12, 13, 138, 142, 158, 161
   Conceptual inquiry, 80
   Congruence, 10, 32, 35, 119, 124, 126, 174, 175, 176, 177
   Consistency, 10, 26, 32, 33, 35, 108, 137, 148, 155, 174
   Cyclic interaction theory, 9, 33, 95, 104, 137, 138, 139, 153, 166, 167, 177, 178, 179, 181, 182, 183
   Decision, 10, 17, 25, 32, 37, 40, 45, 68, 95, 108, 109, 115, 129, 130, 137, 139, 140, 143, 155, 162, 166, 167, 168, 173, 177
   Design-evaluation cycle, 3, 28, 29, 30, 32, 191
   Divide-and-Conquer, 98
   Feedback design, 150
   Freedom, 5, 12, 13, 17, 18, 44
   Gestalt principles, 47, 53, 155
   Goal constrcution, 182–85
   Goal elimination, 185–87
   Goal-to-Action design, 182, 187, 190
   Guidelines. *See* Guidelines
   Interaction Unit scenarios, 109–14
   Navigation design, 162–64
   Participatory, 130, 131, 133, 137
   Pattern library, 146, 154, 155, 164, 170
   Philosophy, 165
   Principles, 4, 25, 26, 28, 83, 119, 163, 168, 192, 213, 217
   Rationale, 13, 30, 32, 145, 173, 195
   Scenario-based, 99, 102, 126, 150, 231
   Screen, 154–64
   Task-Artifact lifecycle, 102
   Task-based design, 2

technology-centered, 1
　User-Centered Design, 2, 25, 26, 28, 29, 32, 93, 129, 237
　Work practices, 80, 83, 94, 96, 97, 98, 99, 102, 103, 104, 106, 109, 110, 116, 124
Design freedom, 5, 12, 13, 17, 18, 44
Design process
　Contextual design, 26, 33
　Eight golden rules, 26
　Interaction design process, 29
　Norman's golden rules, 25
　Putative design process, 28–35, 35, 37, 95, 119, 130, 218, 238
　Usability engineering, 33
Design-evaluation cycle, 3, 28, 29, 30, 32, 191
Diary studies, 84–86
Digital Multimedia Broadcasting, 7, 43
Dilemmas, 26
Direct manipulation, 8, 123, 128, 131, 172
　Interface objects, 28, 123, 124, 128, 130, 132, 139, 140, 144, 154, 155, 169, 170, 172, 176, 178, 179
　Task objects, 123, 124, 128, 144, 170, 175
Display
　Convertible, 6
Divide-and-Conquer, 98
DMB. See Digital Multimedia Broadcasting
DMB phone, 7, 43

E

Earcon, 140, 145, 178
Early adopter, 11, 39
Easy to use, 30, 57, 76, 93, 223
Easy5, 41
Effect-to-Goal design, 182, 183
Effect-to-goal problem, 27, 178, 183, 187, 190, 201
Eight golden rules, 26
Emulator
　Adobe Device Central C3, 164
　Interface Builder, 164
　Macromedia Director, 164
　Openwave, 164
Emulator testing, 225–26
Environment, 73, 86
　Cultural, 55, 59–60, 65, 66, 78, 209, 210, 220
　Learning, 20, 88
　Mobile, 9, 120
　Physical, 55, 56–57, 56, 63, 65, 66, 78, 132, 143, 151, 209, 210, 220
　Profiles, 67, 78, 84, 90, 91
　Social, 55, 57–59, 65, 66, 78, 209, 210, 220
　Task, 123
　Work, 55, 73, 96, 167, 168

Environment profiles documents, 90
Errors
　Constant, 76
　Expert, 109
　Mode error, 179
　Parallax error, 148
　Post-completion error, 182, 183
　Random, 77
　Testing, 230
Ethnography, 82–87, 82
　Cultural probe, 86
　Diary studies, 83, 84–86
　Interview, 83
　Observation, 82
　Participant observation, 86
　Self-reports, 83
　Visual stories, 83
ETIT. See External Task Internal Task
Evaluation
　Activity walkthrough, 208–14
　Cognitive walkthrough, 193–208
　Coherence, 10, 33, 35
　Collective walkthrough, 167–215
　Congruence, 10, 32, 35, 119, 124, 126, 174, 175, 176, 177
　Consistency, 10, 26, 32, 33, 35, 108, 137, 174
　Design-evaluation cycle, 3, 28, 29, 30, 32, 191
　Expert appraisal, 220, 221
　Field test, 228
　Functional commitments consistency, 170
　Guidelines, 168, 171, 173, 213, 218
　Heuristic evaluation, 168–73
　Interaction unit walkthrough, 174–90
　Norman, 25, 26, 28
　Observation, 224
　Procedural consistency, 170
　Questionnaire, 223, 229, 236
　Usability testing, 217–37
　Visual consistency, 170
Experimenter bias, 226
External Task Internal Task, 127, 128
　Direct manipulation, 128
　Work analysis, 128
Extreme programming, 164

F

Fastap, 150
Feature checklist, 124, 125, 128
Feature fatigue, 1, 12, 13
Feedback, 143–46
　Auditory, 148
　Haptic, 36, 46
　Informative, 26, 67

Tactile, 8, 131, 140, 143, 151, 178
Feedback design, 150–51
   Audio feedback, 150
   Other feedback, 151
   Visual feedback, 151
Field studies
   Cultural probe, 86
   Ethnography, 82–87
   Participant observation, 86
Field testing, 228–29
   Laboratory testing, 228
Finger-based interaction, 143, 147, 148, 150
Firefly, 38
Flexibility, 147
Floor effect, 77
Flow experience, 234–37
   Attention, 235
   Control, 235
   Curiosity, 236
   Playfulness, 236
Focus group, 81–82
   Data gathering, 81–82
   Testing, 217
Formal usability testing
   Field testing, 228–29
   Laboratory testing, 224–27
Formative approach, 73, 102, 119
   Critical Incident Method, 87
Foveal, 51
Full browsing, 16
   Cognitive workloads, 15
   Visuo-spatio perception, 52
Functional commitments consistency, 170
Functionality vs. Usability, 12–13

G

Gender, 44, 45, 67
Geo-tagging
   Location-Awareness, 61
Gestalt principles, 47, 53, 155
   Closure, 48, 53
   Good continuation, 48, 53
   Proximity, 48, 53
   Similarity, 48, 53
Gesture
   Mobile Phone, 143
Gesture-based interaction, 143
Giorgio Armani phone, 161
Global Positioning System, 1, 18, 20, 93, 150
   Location-Awareness service, 61, 88
Globe, 22
Goal action matching, 32, 126
Goal constrcution, 182–85

Goal construction
   Misleading cues, 185
   Missing cues, 185
Goal elimination
   Misleading cues, 187
   Missing cues, 187
Goal elimination, 185–87
Goal-to-action problem, 27, 179, 183, 187, 190, 191, 202, 203, 204, 205, 206, 207
GOMS, 32, 108–9, 109, 110, 116
   Keystroke Level Model, 109
GPS. *See* Global Positioning System
GSM, 43
Guidelines, 25, 32, 55, 61, 128, 131, 138, 154, 168, 171
   Evaluation, 168, 171, 173, 213, 218
   Microsoft, 131
   Symbian, 131
   Web design, 25
Gulf of evaluation, 27
Gulf of execution, 27

H

Hawthorne effect, 227, 229, 231
HCI, 26, 27, 32, 54, 65, 75, 82, 95, 106, 108, 121, 167, 172, 173, 179, 192, 217, 218, 237
   Mobile, 41, 228, 237
Heuristic, 168
Heuristic evaluation, 168–73
   Aesthetics, 172
   Consistency and standards, 170
   Error prevention, 172
   Help, 172
   Matching, 169
   User control and freedom, 169
   Visibility, 168
Hidden mode, 143, 181, 190
   Action-to-effect problem, 143, 181, 190
Hierarchical work analysis, 107–8
High fidelity prototype, 33, 153
HTML, 15, *See* Hyper Text Markup Language

I

ICO, 178
Icons
   Conceptual model, 157
   Culture, 36, 42, 59
   Labels, 157, 161
   Metaphor, 143–44
   Object perception, 48, 53, 156
Income, 45
Indian blood donor

Healthcare, 22
Information design
  Cognitive task analysis, 108–14
  Divide-and-Conquer, 107–8
Input
  Gesture, 143
  Speech, 12, 148
  Stylus pen, 4, 6, 147, 148, 154, 159, 161, 194
  T9, 149
Interaction model, 191
  Interaction unit scenarios, 191
Interaction styles
  Direct manipulation, 8, 28, 123, 124, 128, 130, 131, 132, 140, 154, 172
  Drag-and-drop, 8, 140, 178
  Recall-based, 175, 176
  Recognition-based, 52, 139, 175, 176
  Tab-based, 14, 15, 161, 163
  Touch-sensitive, 5, 6, 8, 12, 15, 18, 36, 45, 54, 143, 148, 149, 161, 163, 168
Interaction Unit Model
  Advantage, 116
  Affordance, 112
  Coherence, 10, 33, 35
  Congruence, 10, 32, 35, 119, 124, 126, 174, 175, 176, 177
  Consistency, 10, 35, 137, 174
  Goal reorganization process, 112
  GOMS, 114–17
  HTA, 114–17
  Notation, 112–14
  Other information, 114
  Recall, 114
  Recognition, 115
Interaction unit scenarios, 32, 106, 109, 110, 174
Interaction Unit scenarios, 109–14
  Work model, 109–14
Interaction unit walkthrough, 174–90
  Activity walkthrough, 174
  Cognitive walkthrough, 174
Interface Builder, 164
Interface objects, 28, 123, 124, 128, 130, 132, 139, 140, 144, 154, 155, 169, 170, 172, 176, 178, 179
Internal audience problem, 12, 153, 195
Interview, 78–81
  Bias, 78
  Conceptual inquiry, 80
  Critical Incident Method, 87
  Electronic, 79
  Formal, 79
  Informal, 79
  Rich picture, 60
  Testing, 223, 226, 230

iPod, 48, 54, 124, 156, 163, 164

J

J2ME, 18

K

Keypad
  Chording keyboard, 149
  ChordTap, 149
  Fastap, 150
  Multitap, 149
  QWERTY, 149, 150
  Softkey, 146, 148, 149, 157, 225
  T9, 149
  Twiddler keyboard, 149
  Virtual keyboard, 150
Keystroke Level Model
  Limitations, 109
Keystroke per character, 149
kspc. *See* Keystroke per character

L

Labels
  Icons, 157
Laboratory testing
  Field testing, 228
Lifecycle
  Task-Artifact, 102
  Usability Engineering, 26, 33
Likert scale, 76, 77, 236
  Semantic differential scale, 76, 236
Location awareness
  Finding friends, 61
  Geo-tagging, 61
  GPS, 150
  Mobile learning organizer, 88
  Starbucks coupons, 42
  T-Friends, 42
Low fidelity prototype, 130, 132, 137

M

Macromedia Director, 164
Magnetic sensor, 143
Mapping
  Task-Action, 126, 170
  Task-Function, 33, 94, 95, 96, 104, 119, 121, 124, 126, 128
MBTI, 59
Measures
  Error, 230
  Flow experience, 234–37

Index

Revisit, 230
Time, 230
Time-out, 230–31
Time-series, 231
Memory
  Dementia, 110, 111
  Miller's magic number, 162
  Motor, 39, 147, 149, 170
  Spatial, 14, 54, 55, 160, 170
  Verbal, 14, 55
  Working memory, 53–55
Mental model, 32, 51, 54, 124, 128, 137, 153, 157, 158, 179
  Consistency, 51, 108, 124, 128
Menu design, 158–62
  Alphabetical, 158, 159
  Horizontal, 145, 159, 162, 163
  List-based, 160
  Pop-up, 132, 154, 159
  Pull-down, 154, 159
  Spatial memory, 14, 54, 160
  Tab, 161
  Table-based, 160
Messaging
  Multimedia message, 18, 44
  Text message, 18, 19, 22, 42, 43, 44, 169
Metaphor, 143–44, 143, 154, 157
Misleading mode signal, 143, 181
  Action-to-effect problem, 143
Misleadning mode signal
  Action-to-effect problem, 181
MMS. *See* Messaging
Mnemonics, 26
Mobile business
  Globe, 22
  M-PESA, 22
  WIZZIT, 22
Mobile device
  Blackberry, 42, 132
  MP3 Player, 8
  Personal Multimedia Player, 8
  Treos, 42
Mobile etiquette, 61
Mobile feature
  Communicative, 12
  Personal, 12, 18, 19, 40, 55, 120, 131
  Portable, 4, 12, 16, 19, 20, 40, 55, 238
  Wakable, 12
Mobile ladies, 19
Mobile phone
  Apple iPhone. *See* Apple iPhone
  Blackberry, 42
  Convergent information device, 93

DMB, 7, 43
Easy5, 41
Firefly, 38
Motorola DynaTAC, 11
Nokia 6630, 159
Nokia N70, 4
Samsung Giorgio Armani phone, 161
Samsung Haptic phone, 123, 148
Samsung Motion phone, 143
Sony-Ericsson W21S, 159
Treo, 42
Watch phone, 121
Wrist phone, 121
Mobile Phone
  Gesture, 143
  Samsung SCHS310, 143
Mobile user experience, 234–37
Mode dependency, 181
  Action-to-effect problem, 181
Model
  APT, 190
  Cyclic interaction model, 28
  GOMS, 32, 108
  ICO, 178
  Interaction unit model, 104, 112–14
  Syndetic modeling, 192
  Task-Action Grammar, 32, 108, 170, 171, 172
Model of Human Processor, 46
  Attention, 49–53
  GOMS, 108
  Memory, 53–55
  Visual perception, 46
Motion detection
  Accelerometer, 143
  Magnetic sensor, 143
Motorola DynaTAC, 11
M-PESA, 22
Multitap, 149

N

Navigation design, 162–64
  Affinity analysis, 163
  Card sorting, 163
Neologism, 1
  m-goverment, 1
  m-office, 1
Nintendo DS, 147
Noise
  Interface design, 54, 56, 145
Nokia 6630, 159
Nokia N70, 4
Novice user, 52, 54, 67, 68, 71, 172, 176, 187
NTT DoCoMo, 121

O

Object perception
    Gestalt principles, 47, 155
Observation, 73–75
    Data gathering, 73–75
    Diary studies, 84
    Ethnography, 83
    Obtrusive, 74
    Unobtrusive, 74
    Verbal protocol, 231
Older users, 14, 41, 45, 52, 55, 68
Openwave, 164

P

Palm, 18
Parallax error, 148
Partially hidden mode, 143, 181
    Action-to-effect problem, 143, 181
Participant bias
    Acquiescence effect, 76, 227
    Hawthorne effect, 227, 229, 231
Participant observation, 86
Participatory design, 130, 131, 133, 137
    Sticky note, 131
Participatory evaluation, 221
Pattern library, 146, 154, 155, 164, 170
Perceived affordance, 121
Perception
    Auditory perception, 49
    Gestalt principles, 47, 53, 155
    Visual perception, 46–48, 47, 53, 155
Personal Multimedia Player, 8
Personality, 59
    MBTI, 59
Personalization, 40, 88, 131
    Automatic categorization, 40
Physical affordance, 115, 140
Platform
    BREW, 18
    J2Me, 18
    Palm, 18
    Symbian, 18
    Windows Mobile, 18
    WIPI, 18
Pleasant to use, 235
Post-completion error, 182
Post-test survey, 217
PowerPoint prototype, 134
    Limitations, 134
Preattentive process, 53
Predictive model

GOMS, 108
Privacy, 73, 148, 150
Procedural consistency, 170
Profiles
    Environment profiles, 30, 33, 67, 78, 90, 91
    User profiles, 55, 67, 68, 71, 72, 78, 80, 89, 90, 225
Prototype
    High fidelity, 33, 153
    Low fidelity, 130, 132, 137
    Menu design, 158–62
    PowerPoint, 134
    Sticky-note, 131–33
    Task-Function mapping, 119, 128
    Test-driven, 164
    Video projection, 166
    Wizard-of-Oz, 133
Putative design process, 28–35, 35, 37, 95, 119, 130, 218, 238

Q

Questionnaire, 75–78
    Acquiescence effect, 76
    Ceiling effect, 77
    Data gathering, 75–78
    Floor effect, 77
    Internet, 77
    Likert scale, 76
    Neutral value, 76
    Semantic differential scale, 76
Quick and dirty usability testing, 221
QUIS, 77
    Questionnaire, 77
QWERTY, 149, 150

R

Radio Frequency IDentification, 23
Randomization, 227
    Random assignment, 227
    Random selection, 227
Requirements, 20, 43, 44, 62, 78, 87, 90, 95, 97, 103, 129, 130, 164
    Analysis, 30
    Cognitive, 66, 71, 99
    Context, 33, 55, 63, 83, 88
    Cultural, 36
    Functional, 170, 171
    Gathering, 29, 30
    Older users, 41
    Organizational, 30
Revisit, 230
RFID. *See* Radio Frequency IDentification

Index

Rich picture, 60–61
Rural information helpline, 19

S

Samsung Haptic phone, 123, 148
Samsung Motion phone, 143
Scenario-based design, 99, 102, 126, 150, 231
Screen design, 154–64
Semantic affordance, 115, 128
Semantic differential scale, 76, 236
Seven staged model, 26, 27, 28
Short term memory. *See* Working memory
Simulation
    Adobe Device Central C3, 164
    Macromedia Director, 164
    Openwave, 164
Smart home, 83
SMS. *See* Messaging
Social networking, 61, 209, 210, 211, 212, 213
Softkey, 146, 148, 149, 157, 225
Sony-Ericsson W21S, 159
Standardization
    Always-available connection, 18, 44
    Capability, 18
    Feature, 18
    Fitness to the context, 18
    Form, 18
    User interface, 18
Star user interface, 11
Sticky-note prototype, 131–33
Strong affordance, 182, 183, 187, 202
Super-goal Kill-off, 182
    Post-completion error, 182
Supermarket management system, 23, 38, 39, 56, 58, 94, 99
Symbian, 18
Syndetic modeling, 192
System image, 12, 153, 195

T

T9, 149
    Cognitive dissonance, 149
Tab-based menu design, 161
Talking with existing users, 222
Task
    Classifying, 97, 99
    Decomposing, 98
    Proposing new tasks, 102
    Rearranging, 99
Task completion time, 230
Task objects, 123, 124, 128, 144, 170, 175
Task procedure design, 106, 140–42
    Mobile, 120, 175
Task-Action Grammar
    Consistency, 108, 170
Task-Artifact lifecycle, 102
Task-Function incongruence, 128
Task-Function mapping, 124–29
    Card sorting, 128
    ETIT, 127
    Young's model, 126
Technology Adoption, 43
    Income, 45
Technology development
    Three stages, 11
Technology diary, 85
Technology-centered design, 1
Teens, 44, 120
Test-driven prototype, 164
Testing, 217–37
    Emulator, 225–26
    Field, 228–29
    Post-test survey, 217, 223
    Questionnaire, 223
Text entry, 149–50
Texting. *See* Message:Text messages
    Banking, 22
    Education, 20
    GSM, 43
T-Friends, 42
Think-aloud, 217, 231–34
Third generation, 6, 20, 44, 142, 194
Time-out, 190, 230
Touch-sensitive, 5, 6, 8, 12, 15, 17, 18, 36, 143, 148, 149, 161, 168
Training wheel, 54, 67
Treo, 42
TV
    DMB, 6, 18, 43
Twiddler keyboard, 149

U

Ubiquity, 4
UCD. *See* User-Centered Design
Usability
    Activity walkthrough, 208
    Feature fatigue, 13
    Heuristic evaluation, 168–73
    Interaction unit walkthrough, 174
Usability Engineering Lifecycle, 26, 33
Usability testing, 217–37
    Participatory evaluation, 221
    Quick and dirty usability testing, 221
    Talking with existing users, 222
    Wizard of Oz technique, 222

Usability vs. Functionality, 12–13
Usable, 11
Useful, 11, 16, 30, 93
Useful Field Of View (UFOV), 52
User characteristics
    Age, 35, 41, 44, 67, 68
    Gender, 45
    Income, 45
User experience, 234–37
User model, 2, 80, 137, 176
    Dominance, 176
    Interaction unit model, 177
User profiles, 55, 67, 68, 71, 72, 78, 80, 89, 90, 225
User-Centered Design, 2, 25, 26, 28, 29, 32, 93, 129, 237
Users, 11
    Older users, 14, 41, 45, 52, 55, 57
    Primary, 37, 38
    Secondary, 39
    Teens, 44, 120

V

Verbal protocol, 233
Video projection, 166
Virtual keyboard, 150
Visual consistency, 170
Visual dominance, 49
Visual sampling, 51
Visual scanning
    Pursuit movement, 51
    Saccadic movement, 51

W

Walkthroughs

Activity walkthrough, 208
Cognitive walkthrough, 193–208
Interaction unit walkthrough, 176
Watch phone, 121
Weak affordance, 190
Windows Mobile, 18
WIPI, 18
Wizard of Oz technique, 133, 222
WIZZIT, 22
WML, 15, *See* Wireless Markup Language
Words Per Minute, 149
Work analysis
    Action-effect design, 140
    GOMS, 108
    Hierarchical work analysis, 107–8
    Information design, 158, 163, 167
    Interaction unit model, 109
    Task-Artifact lifecycle, 102
    Task-function mapping, 124, 128, 129
Work context analysis, 36–72
    Activity Theory, 61–66
    Rich picture, 60–61
Workloads
    Cognitive workload, 14, 15, 98, 99, 116, 162, 220, 229
    Designer's workload, 154
wpm. *See* Words Per Minute
Wrist phone, 121

X

XP. *See* Extreme Programming